Engineering fundamentals

R L Timings

Longman
Scientific &
Technical

Longman Scientific & Technical,
Longman Group UK Limited,
Longman House, Burnt Mill, Harlow,
Essex CM20 2JE, England
and Associated Companies throughout the world.

First published 1988
Second impression 1989
Third impression 1990

British Library Cataloguing Publication Data
Timings, R.L.
    Engineer fundamentals.
    1. Engineering
    I. Title
620   TA145

ISBN 0-582-01644-4

Printed in Malaysia
by Ling Wah Press Sdn. Bhd.,
Subang Jaya, Selangor Darul Ehsan

Contents

Acknowledgements

We are indebted to the following for permission to reproduce copyright material:

Extracts from BS 308: Part 1: 1984 (as amended) are reproduced by permission of BSI. Complete copies can be obtained from them at Linford Wood, Milton Keynes, MK14 6LE.

1 Materials

Figure 1.1 shows three objects. The first is a connector joining electric cables. The plastic casing has been partly cut away to show the metal connector. Plastic is used for the outer casing because it is a good electrical insulator and prevents electric shock if a person touches it. It also prevents the connectors from touching each other and causing a short circuit. As well as being a good insulator, the plastic is cheap, tough, and easily moulded to shape. It has been chosen for the casing because of these properties. That is, the properties of cheapness, toughness, good electrical insulation, and ease of moulding to shape.

The metal connector and its clamping screws are made from brass. This metal has been chosen because of its special properties. These properties are: good electrical conductivity, ease of machining to shape, adequate strength, and corrosion resistance. The precious metal silver is an even better conductor of electricity, but it would be far too expensive for this application and it would also be too weak and soft.

The second object in Fig. 1.1 is the connecting rod of a motor car engine. This is made from a special steel alloy. This alloy has been chosen because it combines the properties of strength, toughness, and the ability to be readily forged and machined to shape.

The third object shown in Fig. 1.1 is part of a machine tool. It is the 'tailstock' casting for a small lathe. The metal used in this example is cast iron. The metal has been chosen because it combines the properties of adequate strength with ease of melting and casting to complicated shapes in a simple sand mould. It is also relatively easy to machine to its finished shape.

·Thus the reasons for selecting the materials in the above examples can be summarised as:

(a) Commercial factors such as:
    (i) cost;
    (ii) availability.
(b) Properties of the material.

The properties of the materials can be subdivided further as follows.

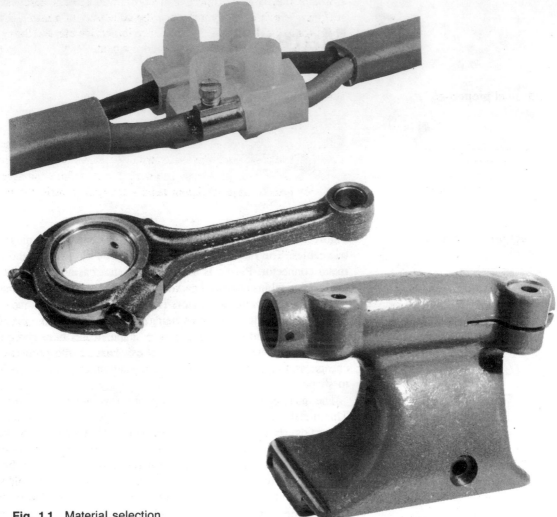

Fig. 1.1 Material selection

Chemical properties
These are the ways in which any given substance behaves in the presence of any other substance. For example, the corrosion resistance of a metal in a given environment, or the resistance of a plastic to a particular solvent.

Physical properties
These are such properties as the density of a material, the electrical and thermal conductivity of a metal, or the electrical and thermal insulating properties of a plastic. They may include such important properties as the melting point of a material (fusibility) or the ability of a material to withstand high temperatures (refractoriness). The magnetic properties of a material are also included in this category,

as are the mechanical properties of strength, toughness, hardness and stiffness, which not only influence the suitability of a material for a given component but will also influence its service life, and the choice of manufacturing process for that component.

1.2 Physical properties

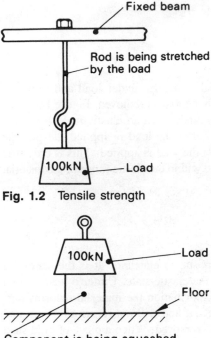

Fig. 1.2 Tensile strength

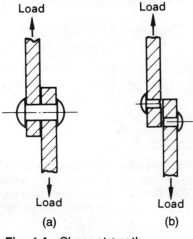

Component is being squashed by the load

Fig. 1.3 Compressive strength

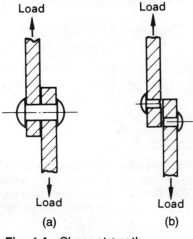

Fig. 1.4 Shear strength

Tensile strength

This is the ability of a material to withstand tensile (stretching) loads without breaking. Figure 1.2 shows a heavy load being held up by a rod fastened to a beam. The load is trying to *stretch* the rod. Therefore the rod is said to be in *tension*, so the material from which the rod is made needs to have sufficient *tensile strength* to resist the pull of the load.

Compressive strength

This is the ability of a material to withstand compressive (squeezing) loads without being crushed or broken. Figure 1.3 shows a component being crushed by a heavy load. The load is said to be *compressing* the component. Therefore the component needs to be made from a material with sufficient *compressive strength* to resist the load.

Shear strength

This is the ability of a material to withstand offset loads, or transverse cutting (shearing actions). Figure 1.4 shows a rivet joining two metal bars together. The forces acting on the two bars are trying to pull them apart (a). Because the loads are not exactly in line, they are said to be *off-set* and, therefore, the load on the rivet is called a *shearing* load. That is, the rivet is said to be *in shear*. If the rivet material does not have sufficient *shear strength* to resist the loads, the rivet will break (shear-off) as shown and the bars and the loads acting upon them will move apart as shown (b). The same effect can be caused by loads pushing on the ends of the two metal bars joined by the rivet.

Toughness (Impact Resistance)

This is the ability of a material to withstand shatter. If a material shatters it is *brittle* (e.g. glass). Rubber and most plastic materials do not shatter, therefore they are *tough*. Toughness should not be confused with strength. Figure 1.5 shows a metal rod being broken by impact loading in a vice. If the metal is, say, a piece of high carbon steel rod (e.g. silver steel) in the annealed (soft) condition as supplied, it will have only a moderate tensile strength but under the impact of the

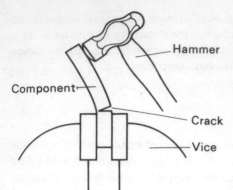

Fig. 1.5   Impact resistance

hammer it will bend without breaking and is, therefore, *tough*. If a similar specimen is made hard by making it red-hot and cooling it quickly in water, it will now have a very much higher tensile strength. However, although it is now *stronger* it will prove to be brittle and break off easily when struck with a hammer and therefore now *lacks toughness*. A material is brittle and shatters because small cracks in its surface grow quickly under a tensile force. Therefore any material in which the spread of surface cracks does not occur or only occurs slowly and to a small extent is said to be *tough*.

Elasticity

This is the ability of a material to deform under load and return to its original size and shape when the load is removed. Figure 1.6 shows a tensile test specimen. If it is made from an elastic material it will be the same length before and after the load is applied, despite the fact that it will be longer whilst the load is applied. This is only true if the load is relatively small and within the elastic range of the material being tested.

Plasticity

This property is the exact opposite to elasticity. It is the state of a material when loaded beyond the elastic state. Under a load beyond that required to cause elastic deformation the material deforms permanently, it takes a *permanent set*, and will not return to its original size and shape when the load is removed. When a piece of mild steel is bent at right-angles into the shape of a bracket it shows the property of plasticity since it does not spring back straight again. This is shown in Fig. 1.7. *Ductility* and *malleability* are particular cases of the property of plasticity and they will now be considered separately.

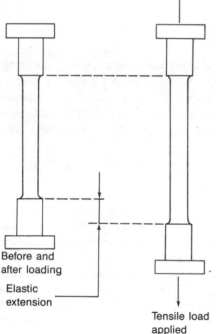

Before and
after loading

Elastic
extension

Tensile load
applied

Fig. 1.6   Elasticity

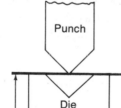

Punch

Die

Strip before bending
force is applied

(i)

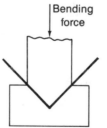

Bending
force

Strip bent beyond the
elastic limit so that
plastic deformation
occurs

(ii)

Since plastic flow has
occurred during bending
strip remains bent after
the bending force has
been removed

(iii)

Fig. 1.7   Plasticity

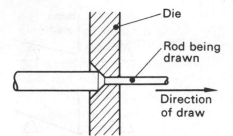

A rod being drawn through a die to reduce its diameter requires the property of ductility

Fig. 1.8 Ductility

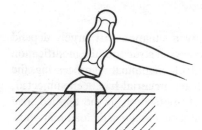

Forming the head of a rivet by hammering. The rivet needs to be made from a malleable material to withstand this treatment

Fig. 1.9 Malleability

Fig. 1.10 Hardness

Ductility

This is the term used when plastic deformation occurs as the result of applying a tensile load. A ductile material is required for such processes as wire drawing (Fig. 1.8), tube drawing and pressing out motor car body panels.

Malleability

This is the term used when plastic deformation occurs as the result of applying a compressive load. A malleable material is required for such processes as forging, rolling, and rivet heading (Fig. 1.9).

Hardness

This is defined as the ability of a material to withstand scratching (abrasion) or indentation by another hard body. It is an indication of the wear resistance of the material. Figure 1.10 shows a hardened steel ball being pressed first into hard material and then into a soft material by the same load. The ball only makes a small indentation in the hard material, but it makes a deeper indentation in the soft material. Hardness is usually tested in this manner, the depth of indentation or the area of the impression being taken as an indication of the hardness of the material.

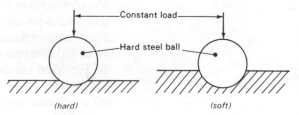

Rigidity (stiffness)

This is a measure of a material's ability not to deflect when subjected to an applied load. For example, although steel is stronger than cast iron, the latter material is preferred for machine beds and frames because it is more rigid and less likely to deflect with consequent loss of alignment and accuracy. Consider Fig. 1.11. For a given load the cast iron bar deflects less than the steel bar because cast iron is a more rigid material. However when the load is increased (Fig. 1.11(b)) the cast iron bar will break, whilst the steel bar merely deflects a little further but does not break. Thus a material which is rigid is not necessarily strong. In fact, as has been shown, the opposite is more often true since rigid materials are often brittle.

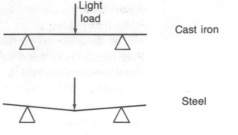

Cast iron

Steel

Under a light load cast iron
deflects less than steel since
cast iron is more rigid

Under a heavy load the cast
iron breaks whilst the
stronger but less rigid steel
merely bends further

(a)   (b)

Fig. 1.11   Rigidity

The response of a material to a given situation will largely depend
upon its chemical composition and crystal structure, the modification
this basic structure has received during manipulative processing, the
heat treatment processes to which the material has been subjected,
and the magnitude and method of application of the applied load
(force). For instance, a material subjected to a tensile load may show
a markedly higher strength when the load is applied slowly than when
the load is applied suddenly. Again, a material such as mild steel will
show elastic properties when only a small load is applied, yet it will
show plastic properties when a larger load is applied. Again, a
heterogeneous material such as cast iron may show a high strength
when subjected to compressive loading, but show a low strength when
subjected to tensile loading. Therefore when choosing a material for
a particular application, or when choosing a manufacturing process
for a particular material, great care should be taken when interpreting
the published test data. Allowance must be made for the difference
between the conditions under which the test data was obtained and
the conditions under which the material will be loaded during processing or in service.

Electrical conductivity

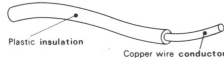

Plastic insulation

Copper wire conductor

Fig. 1.12   Electrical conductivity

Figure 1.12 shows a piece of electrical cable. In this example copper
wire has been chosen for the central conductor or core of the cable
because copper has the property of very good *electrical conductivity*,
that is, if offers a very low resistance to the flow of electrons (electric
current) along the wire. A plastic material such as polymerised vinyl
chloride (PVC) has been chosen for the insulating sheathing surrounding the wire conductor. This material has been chosen because it is
such a bad conductor that very few electrons can pass through it. Very
bad conductors such as PVC are called *insulators*. There is no such
thing as a perfect insulator material, only extremely bad conductors.

Again, as when discussing mechanical properties, care must be taken when interpreting tables of test data. For example, metallic conductors of electricity all increase in resistance as their temperature rises. Pure metals show this effect more strongly than alloys. However, pure metals generally have a better conductivity than alloys at room temperatures. The conductivity of metals and metallic alloys improves as their temperature falls. Conversely, non-metallic materials used for insulators tend to offer a lower resistance to the flow of electrons and so become poorer insulators as their temperatures rise. Thus care must be taken if cables are bunched together without adequate ventilation. Glass, for example, is an excellent insulator at room temperature, but becomes a conductor if raised to red-heat. Table 1.1 lists the properties of typical conductor materials and Table 1.2 lists the properties of typical insulating materials.

Magnetic conductivity

Just as some materials can be good or bad conductors of electricity, so some materials can be good or bad conductors of magnetism. The good magnetic conductors are the 'ferro-magnetic' materials which get their name from the fact that they are made from irons, steels and associated alloying elements such as cobalt and nickel. All other materials are non-magnetic and offer a high *reluctance* (resistance) to the magnetic flux field.

Magnetic materials can be classified as either hard or soft. *Hard magnetic materials* retain their magnetism after the initial magnetising force has been removed. Traditionally, permanent magnets were made from quench-hardened high carbon steels. However, modern permanent magnet alloys are more powerful and are used for all but the most simple, low cost applications. All these materials get their name from the fact that they are very hard and cannot be machined except by grinding. Table 1.3 lists some typical ferro-magnetic alloys for permanent magnets and their corresponding BH values. The BH_{max} value is the maximum magnetic energy which the magnet can give out. It can be seen that an alloy such as 'Columax' is nearly 34 times more powerful than a similar magnet made from high carbon steel. However, these materials have a low *permeability* and are relatively difficult to magnetise but retain their magnetism well once they have been magnetised.

Soft magnetic materials are the opposite of permanent magnetic materials and they have a high *permeability*. This means that they are very easily magnetised and demagnetised and that they retain virtually no magnetism when the magnetising force is removed. Traditionally low-carbon steels and wrought iron were used for transformer, choke and electromagnet cores and for motor and generator pole pieces. Nowadays, however, these materials have been superseded by special alloys possessing very low hysteresis values and very high permeability.

Table 1.1 Properties of conductor materials

Material	Resistivity ($\mu\Omega$ mm)	Temperature coefficient α_0 (per °C)
Aluminium	28.0	42×10^{-4}
Carbon (graphitic)	46.0×10^3	-5×10^{-4}
Copper (annealed)	17.2	43×10^{-4}
Mild steel	107.0	65×10^{-4}
Manganin alloy (1)	480.0	0
Nichrome alloy (2)	1090.0	53×10^{-4}
Nickel	136.0	56×10^{-4}
Silver (annealed)	15.8	41×10^{-4}

Note: (1) *Manganin* is an alloy of copper, manganese and nickel and is used for wire wound resistors for use in measuring instruments where its zero value of temperature coefficient means that the ohmic value of such resistors are unaffected by temperature change.

(2) *Nichrome* is an alloy of nickel and chromium which resists oxidation at high temperatures. It is used for heating elements in electric radiators and furnaces.

EXAMPLE 1

Calculate the resistance of 50 m of copper wire having a cross-sectional area of 0.05 mm². (ρ for Cu = 17.2 $\mu\Omega$mm)

$$R = \frac{\rho l}{a}$$

where: $\rho = 17.2\ \mu\Omega$mm
$l = 50 \times 10^3$ mm
$a = 0.05$ mm²

$$= \frac{17.2 \times 10^{-4} \times 50 \times 10^3}{0.05}$$

$$= \underline{17.2\ \Omega}$$

EXAMPLE 2

A solenoid wound from copper wire draws 1.5A from a 240 V supply at 15°C. Calculate the current when the temperature of the solenoid rises to 32°C ($\alpha_0 = 43 \times 10^{-4}$/°C)

$$R_1 = \frac{V}{I} = \frac{240V}{1.5A} = 160\ \Omega$$

$$\frac{R_1}{R_2} = \frac{1 + \alpha_0\theta_1}{1 + \alpha_0\theta_2}$$

where: $R_1 = 160\ \Omega$
$R_2 = $ resistance at 32°C
$\theta_1 = 15°C$
$\theta_2 = 32°C$
$\alpha_0 = 43 \times 10^{-4}$/°C
$I = $ current

$$\frac{160}{R_2} = \frac{1 + (43 \times 10^{-4} \times 15)}{1 + (43 \times 10^{-4} \times 32)}$$

$$\frac{160}{R_2} = \frac{1.0645}{1.1376}$$

$$R_2 = \frac{160 \times 1.1376}{1.0645}$$

$$R_2 = 171\ \Omega$$

$$I = \frac{V}{R} = \frac{240}{171} = \underline{1.4A}$$

Table 1.2 Properties of insulating materials

Material	Typical applications	Properties
Insulating oil	Switch gear, transformers, cable impregnation	Reduces arcing between switchgear contacts. Being a fluid insulator it can circulate by convection and cool the winchings and cores of large transformers as well as insulating them. Used to impregnate the paper insulation in underground armoured mains cables.
Paper	Armoured mains cables	A good, relatively cheap, insulator for rigid cables. Must be oil impregnated and sealed against ingress of moisture which causes the oil to break down.
Rubber (natural)	Flexible cables	Flexible, high insulation resistance, reasonable mechanical properties when 'vulcanised' with 5% sulphur. Degrades rapidly (perishes) in strong sunlight and undue heat (max 55°C). Sulphur content attacks copper conductors which, therefore, must be tinned.
Silicone rubber (synthetic)	Flexible cables	Similar to rubber, but suitable for applications up to 150°C, very much more expensive.
Polyvinylchloride (PVC)	Flexible cables	Although PVC has a much lower insulation resistance than rubber it is now more widely used because of its low cost and resistance to oils, petrol and chemical solvents. Max operating temperature 65°C.
Mineral insulation (magnesium oxide)	Rigid metal sheathed (MIMS) cables sheathed heating elements	Can operate at temperatures up to dull-red heat. Limited to 660 Volts. Terminations must be sealed as magnesium oxide powder is hygroscopic.
Thermosetting plastics	Moulded insulators (interior)	(i) Phenolic resins (Bakelite) used for moulded insulation blocks and switchgear components where strength is important and its dark colour acceptable. (ii) Amino resins are used for domestic switchgear mouldings as it is available in white and light colours. Lower strength than phenolic resin.
Glass & ceramics	Moulded insulators (exterior)	Hard, highly glazed surface, prevents weathering. Weak in tension and shear, insulators must be designed to operate under compressive mechanical loads only. Used for high voltage insulators for overhead transmission lines. High insulation resistance. Woven glass fibre (resin varnish impregnated) used for high temperature insulation and sheathing in domestic cookers.

Table 1.3 Permanent magnet materials

Name	Composition (%)										Magnetic properties		
	C	Cr	W	Co	Al	Ni	Cu	Nb	Ti	Fe	B_{rem} (T)	Hc (A/m)	BH_{max}
Quench hardened High-carbon steel	1.0	—	—	—	—	—	—	—	—	Rem	0.9	4 400	1 560
35% cobalt steel	0.9	6.0	5.0	35.0	—	—	—	—	—	Rem	0.9	20 000	7 800
Alnico				12.0	9.5	17.0	5.0	—	—	Rem	0.73	44 500	13 500
Alcomax III*				24.5	8.0	13.5	3.0	0.6	—	Rem	1.26	51 700	38 000
Hycomax III*				34.0	7.0	15.0	4.0	—	5.0	Rem	0.88	115 400	35 200
Columax**				24.5	8.0	13.5	3.0	0.6	—	Rem	1.35	58 800	52 800

* Anisopropic alloys whose magnetic properties are measured along the preferred axis
** This alloy derives its very high BH value from the way it is cooled during casting which orientates its columnar crystals parallel to the preferred axis of magnetisation
C = carbon   Cr = chromium   W = tungsten   Co = cobalt   Al = aluminium   Ni = nickel
Cu = copper   Nb = niobium   Ti = titanium   Fe = iron

Two such alloys are:

(a) *Mumetal.* This is an alloy containing 74% nickel, 5% copper, 1% manganese, and 20% iron. This material is widely used as a shielding material in telecommunications equipment.

(b) *Silicon-iron.* This contains 4.0% silicon, 0.3% manganese, less than 0.05% carbon, and the remainder iron. This alloy is very much cheaper than mumetal whilst still possessing very low hysteresis values. It is widely used for lamination stampings for transformers, chokes, electromagnets, and motor and generator rotors and stators.

Thermal conductivity

This is the ability of a material to transmit heat energy by conduction. Figure 1.13 shows a soldering iron. The *bit* is made from copper which is a good conductor of heat and so will allow the heat energy

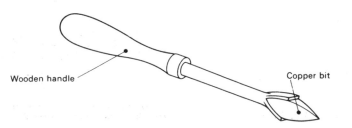

Wooden handle

Copper bit

Fig. 1.13 Thermal conductivity

Table 1.4 Thermal properties of materials

Material	Melting point	Conductivity
Aluminium	660°C	Very good
Copper	1080°C	Excellent
Iron	1535°C	Good
Wood	No melting point (burns)	Poor (good insulator)
Polystyrene (expanded)	No defined melting point but softens at 100°C	Very Poor (Excellent insulator)
Glass Fibre	No defined melting point but softens at about 600°C to 800°C depending on composition	Very poor (very good insulator)
Fire bricks & clays for furnace linings	1595°C to 1800°C depending upon alumina content softening and loss of strength is progressive and commences below these temperatures	Poor (good insulator)

stored in it to travel easily down to the tip and into the work being soldered. The wooden handle remains cool as it has a low thermal conductivity and resists the flow of heat energy. Table 1.4 lists the thermal properties of a number of typical engineering materials.

Fusibility

This is the ease with which materials will melt. It is shown in Fig. 1.14 that solder melts easily and so has the property of *high fusibility*. On the other hand, fire brick only melts at very high temperatures and so has the property of *low fusibility*. Materials which will only melt at very high temperatures are called *refractory materials*. These must not be confused with materials which have a low thermal conductivity and are used as thermal insulators. Reference to Table 1.4 shows that although expanded polystyrene is an excellent thermal insulator, it has a very low melting point.

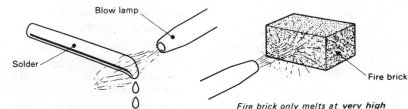

Fig. 1.14 Fusibility

Solder melts at a **low temperature** Therefore it has a **high fusibility**

Fire brick only melts at **very high temperatures** Therefore it has a **low fusibility** and is called a **refractory**

Temperature stability

There are two effects which changes in temperature can have on a component.

(1) *Thermal expansion.* All materials, to a greater or lesser extent, expand when heated and contract when cooled. This expansion and contraction is proportional to the change in temperature and is dealt with in depth in the *Engineering Science* unit.

(2) *Creep.* This is an important factor when considering polymeric (plastic) materials. It must also be considered when metals are expected to work continuously at high temperatures, for example gas-turbine blades. Creep is defined as the gradual extension of a material over a long period of time whilst the applied load is kept constant. Figure 1.15 shows some typical creep curves for cellulose acetate at 25 °C. It can be seen that the creep rate is highest for large applied loads. The creep rate would be greater if the temperature was raised and would be less if the temperature was lowered.

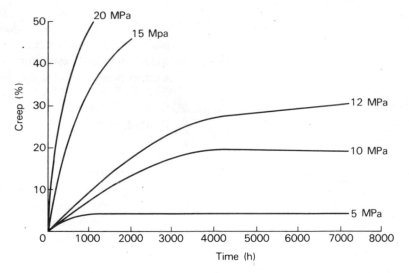

Fig. 1.15 Creep (cellulose acetate)

1.3 Ferrous metals

Ferrous metals and alloys are based upon the metal iron which is their main constituent. They get their name from the Latin word for iron which is *ferrum*. Iron is a soft grey metal rarely found in the pure state outside the laboratory. The engineer usually finds it alloyed, or associated, with the non-metal *carbon*. Coke is a form of carbon, and it is from the coke used in blast furnaces to extract the iron from its ore that the iron-carbon compounds are initially formed. This association with carbon greatly modifies the behaviour of the iron, making it harder, stronger, and of greater use to the engineer. Slight variations in the amount of carbon present can make very great differences in the properties of the metal. Table 1.5 shows how the addition of

Table 1.5 Ferrous metals

Name	Group	Carbon content %	Some uses
Dead mild steel	Plain carbon steel	0.1 to 0.15	Sheet for pressing out such shapes as motor car body panels. Thin wire, rod, and drawn tubes
Mild steel	Plain carbon steel	0.15 to 0.3	General purpose workshop bars, boiler plate, girders
Medium carbon steel	Plain carbon steel	0.2 to 0.5 0.5 to 0.8	Crankshaft forgings, axles Leaf springs, cold chisels
High carbon steel	Plain carbon steel	0.8 to 1.0 1.0 to 1.2 1.2 to 1.4	Coil springs, wood chisels Files, drills, taps and dies Fine-edge tools (knives, etc.)
Grey cast iron	Cast iron	3.2 to 3.5	Machine castings

varying amounts of carbon to the metal iron can produce a wide range of ferrous metals.

Figure 1.16 shows the effect of the carbon content upon the properties of plain carbon steels up to a maximum of 1.2% carbon. The maximum amount of carbon which will combine with iron to form iron-carbide at room temperature is 1.7% but, in actual practice, there is little or no advantage in increasing the amount of carbon present above 1.2% and there is always a possibility that as the amount of carbon approaches the maximum some carbon may precipitate out destroying the properties of the steel.

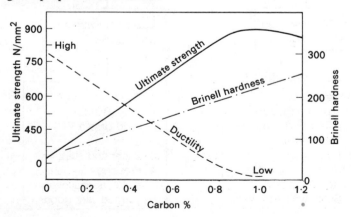

Fig. 1.16 The effect of carbon content on the properties of plain carbon steels (annealed)

1.4   Plain carbon steels

Plain carbon steels are defined as alloys of iron and carbon in which the iron and carbon are combined at all times. Only the range of ferrous metals with their carbon contents lying between the theoretical limits of 0.1% and 1.7% satisfy this definition. In practice an upper limit of 1.2% carbon is rarely exceeded.

Dead mild steel

In dead mild steel the carbon content is deliberately kept as low as possible so that the steel will have a high ductility. This enables it to be pressed into complicated shapes, such as motor car body panels, even whilst it is cold. It is slightly weaker than mild steel and is not usually machined since its softness would cause it to tear and leave a poor finish.

Mild steel

This is a widely used material which is relatively cheap and freely available. It is soft and ductile and can be forged, pressed and drawn in the hot or cold condition. It is easily machined using high-speed steel cutting tools. Typical applications are listed in Table 1.6.

Table 1.6   Mild steel

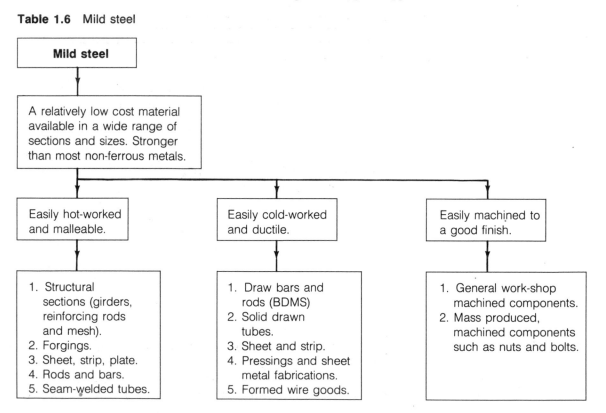

Medium carbon steel

This is harder, tougher and less ductile than mild steel, and cannot be bent or formed in the cold condition to any great extent without the risk of it cracking. However, it hot forges well but close temperature control is required to prevent:

(a) 'burning' at high temperatures (over 1150 °C) which leads to embrittlement;

(b) cracking when forged below 700 °C, due to work hardening.

Typical applications are listed in Table 1.7.

Table 1.7 Medium carbon steel

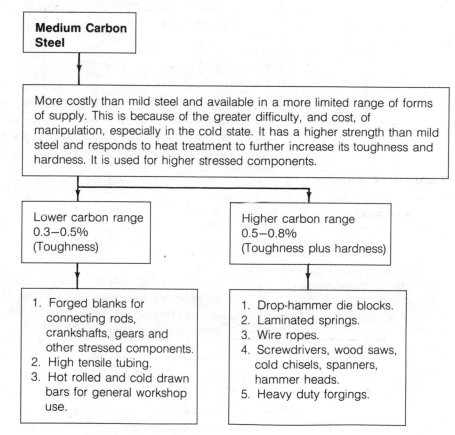

High carbon steel

This is harder, less ductile and slightly less tough than medium carbon steel. Cold forming is not recommended, but it can be hot forged providing the temperature is closely controlled to between an upper limit of 900 °C and a lower limit of 700 °C. High carbon steel is widely

Table 1.8 High carbon steel

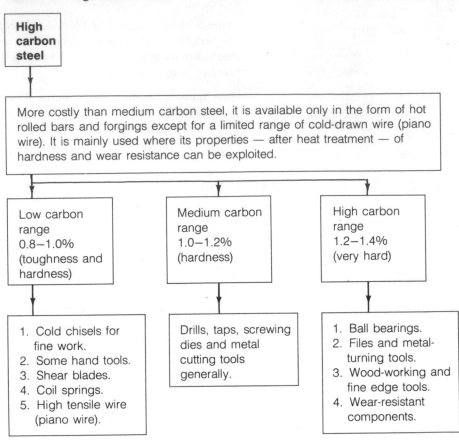

used for cutting tools for hand processes such as filing. For machining processes high-speed steels are used in preference to high carbon steels as they do not soften so easily when they become hot. Typical applications are listed in Table 1.8.

1.5 Grey cast iron

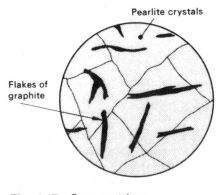

Fig. 1.17 Grey cast iron

Grey cast iron is very similar in composition and properties to the crude pig iron produced by the blast furnace. It does not require the complex and costly refinement processes of steels and, therefore, provides a useful low-cost engineering material (see Table 1.9). Cast irons contain substantially more than the 1.7% carbon that forms the upper limit for plain carbon steels. In fact, the distinguishing characteristic of cast irons is their uncombined or *'free' carbon* content. In grey cast iron the free carbon appears as flakes of graphite as shown in Fig. 1.17. It is these flakes of graphite which give grey cast iron its characteristic colour when fractured, its 'dirtiness' when machined and its weakness when subjected to a tensile load. The graphite also promotes good machining characteristics by acting as an internal lubricant and also producing an easily disposable discontinuous chip. The

Table **1.9** Grey cast iron

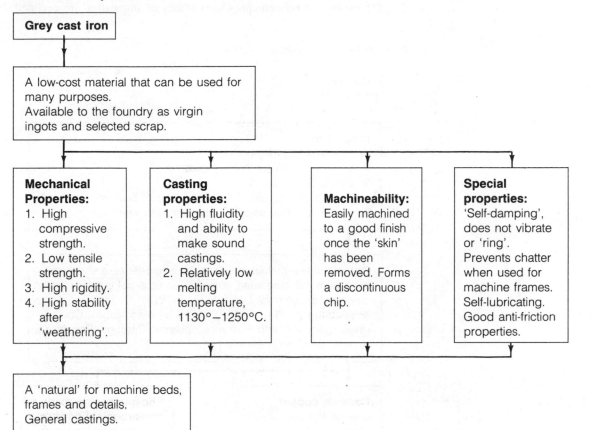

cavities containing the flake graphite have a dampening effect upon vibrations — cast iron is *non-resonant* — and this property makes it particularly suitable for machine tool frames and beds.

1.6 Copper

The 'non-ferrous' metals are all those thirty-eight metals, other than iron, which are known to man. For engineering purposes, copper is one of the most important of these. It is used as the basis for a wide range of brass and bronze alloys. It is widely used for electrical conductors and for heat-exchangers such as motor car radiators. Its main properties can be listed as:

Density	8900 kg/m^3;
Melting point	1083 °C;
Tensile strength	232 MPa
General	Soft, very ductile, relatively low tensile strength, second only to silver in conductivity, it is easy to join by soldering and brazing, it is highly corrosion resistant.

The properties of 'pure' copper depend upon the degree of purity and the method of refinement. Often traces of impurities are retained

Table 1.10 Copper

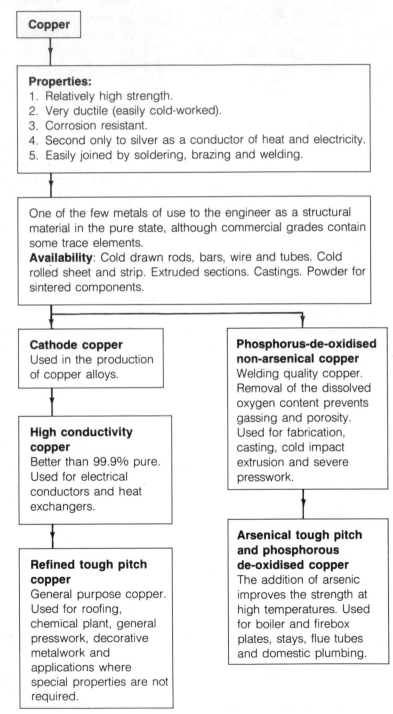

Copper

Properties:
1. Relatively high strength.
2. Very ductile (easily cold-worked).
3. Corrosion resistant.
4. Second only to silver as a conductor of heat and electricity.
5. Easily joined by soldering, brazing and welding.

One of the few metals of use to the engineer as a structural material in the pure state, although commercial grades contain some trace elements.
Availability: Cold drawn rods, bars, wire and tubes. Cold rolled sheet and strip. Extruded sections. Castings. Powder for sintered components.

Cathode copper
Used in the production of copper alloys.

Phosphorus-de-oxidised non-arsenical copper
Welding quality copper. Removal of the dissolved oxygen content prevents gassing and porosity. Used for fabrication, casting, cold impact extrusion and severe presswork.

High conductivity copper
Better than 99.9% pure. Used for electrical conductors and heat exchangers.

Arsenical tough pitch and phosphorous de-oxidised copper
The addition of arsenic improves the strength at high temperatures. Used for boiler and firebox plates, stays, flue tubes and domestic plumbing.

Refined tough pitch copper
General purpose copper. Used for roofing, chemical plant, general presswork, decorative metalwork and applications where special properties are not required.

deliberately to enhance the properties of copper for a particular application (see Table 1.10).

1.7 Copper alloys

Four groups of copper alloys will now be considered:

(a) the brass alloys;
(b) the tin-bronze alloys;
(c) the aluminium bronze alloys;
(d) the cupro-nickel alloys.

Brass alloys

These are alloys of copper and zinc. They tend to give rather poor quality, porous castings and depend upon hot- or cold-working (e.g. rolling, stamping, extrusion, etc.) to consolidate the metal and improve its mechanical properties. The more common brasses are considered in Table 1.11.

Table 1.11 Brass alloys

Name	Composition (%)			Applications
	Copper	Zinc	Other elements	
Cartridge brass	70	30	—	Most ductile of the copper–zinc alloys. Widely used in sheet metal pressing for severe deep drawing operations. Originally developed for making cartridge cases, hence its name
Standard brass	65	35	—	Cheaper than cartridge brass and rather less ductile. Suitable for most engineering processes
Basis brass	63	37	—	The cheapest of the cold working brasses. It lacks ductility and is only capable of withstanding simple forming operations
Muntz metal	60	40	—	Not suitable for cold-working, but hot-works well. Relatively cheap due to its high zinc content, it is widely used for extrusion and hot-stamping processes
Free-cutting brass	58	39	3% lead	Not suitable for cold-working, but excellent for hot-working and high-speed machining of low strength components
Admiralty brass	70	29	1% tin	This is virtually cartridge brass plus a little tin to prevent corrosion in the presence of salt water
Naval brass	62	37	1% tin	This is virtually Muntz metal plus a little tin to prevent corrosion in the presence of salt water

Tin-bronze alloys

These are alloys of copper and tin together with a de-oxidiser. The de-oxidiser is essential to prevent the tin content oxidising during casting and hot-working. Oxidation of the tin would result in a weak, brittle, 'scratchy' bronze. Two de-oxidisers are commonly used:

(*a*) phosphorus in the *'phosphor bronze'* alloys;
(*b*) zinc in the *'gun metal'* alloys.

Unlike the brasses which are largely used in the wrought condition (rod, sheet, extruded sections, etc.), only low tin content bronzes can be worked and most bronze components are in the form of castings. Tin bronzes are more expensive than the brasses, but are stronger and

Table 1.12   Tin-bronze alloys

Name	Composition (%)				Application
	Copper	Zinc	Phosphorous	Tin	
Low-tin bronze	96	—	0.1 to 0.25	3.9 to 3.75	This alloy can be severely cold-worked to harden it so that it can be used for springs where good elastic properties must be combined with corrosion resistance, fatigues resistance and electrical conductivity, e.g. contact blades
Drawn phosphor-bronze	94	—	0.1 to 0.5	5.9 to 5.5	This alloy is used in the work-hardened condition for turned components requiring strength resistance, such as valve spindles
Cast phosphor-bronze	90	—	0.03 to 0.25	10	Usually cast into rods and tubes for making bearing bushes and worm wheels. It has excellent anti-friction properties
Admiralty gunmetal	88	2	—	10	This alloy is suitable for sand casting where fine-grained, pressure-tight components such as pump and valve bodies are required
Leaded-gunmetal (free-cutting)	85	5 (5% lead)	—	5	Also known as 'red brass', this alloy is used for the same purposes as standard, Admiralty gunmetal. It is rather less strong but has improved pressure tightness and machining properties
Leaded (plastic) bronze	74	(24% lead)	—	2	This alloy is used for lightly loaded bearings where alignment is difficult. Due to its softness, bearings made from this alloy 'bed in' easily

give sound, pressure-tight castings which are widely used for steam and hydraulic valve bodies and mechanisms. They are highly resistant to wear and corrosion. Some typical tin bronze alloys are listed in Table 1.12.

Aluminium-bronze alloys

These are more expensive than the 'tin bronze' alloys, but they are more corrosion resistant at high temperatures. They are also more ductile and can, for example, be cold-worked into tubes for boilers and condensers in steam and chemical plant. Typical compositions for these alloys would be:

(a) Wrought alloy

Aluminium	5%
Nickel/Manganese	4%
Copper	remainder

(b) Casting alloy

Aluminium	9.5%
Iron	2.5%
Nickel	1.0%
Manganese	1.0%
Copper	remainder

Cupro-nickel alloys

These are expensive copper based alloys which have special properties where extremes of strength and corrosion resistance are required together with high ductility. They are used for such applications as very high duty boiler and condenser tubes in steam plant, bullet envelopes, and resistance wires. 'Monel metal' has exceptionally high corrosion resistance and is widely used in chemical plant. Cupro-nickel alloys are increasingly used in marine applications because of their superior corrosion resistance. Typical compositions for these alloys would be:

(a) 'Coinage' bronze

Nickel	25.00%
Manganese	00.25%
Copper	remainder

(b) Monel metal

Nickel	68.00%
Iron	1.25%
Manganese	1.25%
Copper	remainder

1.8 Aluminium

Along with copper, aluminium is one of the most widely used non-ferrous metals. As a pure metal it lacks the strength for a structural material, but it is widely used where its low density, resistance to corrosion, and high electrical and thermal conductivity can be exploited. It is used, for example, for the overhead transmission lines of the electricity supply grid system. These conductors consist of a core of very high tensile steel cable to carry the mechanical loads of the long spans used. This core is then surrounded by the aluminium conductors which carry the electricity. The main draw-back to using aluminium for electrical conductors is the difficulty in soldering and brazing it, although this is becoming easier as special fluxes and spelters are developed. For most engineering purposes aluminium is of importance as the basis of a very wide range of important alloys. Its main properties can be listed as:

Density	2700 kg/m^3
Melting point	660 °C
Tensile strength	93 MPa
General	Lightest of the commonly used metals, high electrical and thermal conductivity, soft and ductile.

1.9 Aluminium alloys

Aluminium alloys cover a vast range of engineering materials sold under a bewildering number of proprietary trade names. However, for general engineering purposes, the alloys are listed under BS 1470–77 for wrought metals and under BS 1490 for cast metals and ingots. Airframe and aircraft engine materials are covered by the 'L' series of British Standards and the DTD specifications. Aluminium alloys can be classified into four groups:

(a) wrought alloys (not heat-treatable);
(b) cast alloys (not heat-treatable);
(c) wrought alloys (heat-treatable);
(d) cast alloys (heat-treatable).

Examples of these groups of alloys are listed in Table 1.13. The heat treatment of aluminium alloys is discussed in section 1.17.

1.10 Tin-Lead alloys

Tin and lead form a continuous range of alloys from nearly pure lead and no tin, to nearly pure tin and no lead. The most useful alloys are those ranging from 33% tin and 67% lead (plumber's solder) to 62% tin and 38% lead (tinman's solder). In addition, tin-lead alloys form the basis of many bearing metals and these will be considered in section 1.20. Table 1.14 lists some typical tin-lead solders together with their solidification range and applications.

1.11 Common forms of supply

There is an almost unlimited range to the forms in which metal can be supplied to the user. Figure 1.18 shows some of these forms. The processes by which the metal is produced will have a profound effect upon its properties. Tables 1.15, 1.16 and 1.17 compare the advantages

Table 1.13 Some typical aluminium alloys

Composition (%) (Only elements other than aluminium are shown)						Category	Applications
Copper	Silicon	Iron	Manganese	Magnesium	Other elements		
0.1 max	0.5 max	0.7 max	0.1 max	—	—	Wrought Not heat-treatable	Fabricated assemblies. Electrical conductors. Food and brewing processing plant. Architectural decoration
0.15 max	0.6 max	0.75 max	1.0 max	4.5 to 5.5	0.5 Chromium	Wrought Not heat-treatable	High-strength shipbuilding and engineering products. Good corrosion resistance
1.6	10.0	—	—	—	—	Cast Not heat-treatable	General purpose alloy for moderately stressed pressure die-castings
—	10.0 to 13.0	—	—	—	—	Cast Not heat-treatable	One of the most widely used alloys. Suitable for sand, gravity and pressure die-casting. Excellent foundry characteristics for large marine, automotive and general engineering castings
4.2	0.7	0.7	0.7	0.7	0.3 Titanium (optional)	Wrought Heat-treatable	Traditional 'Duralumin' general machining alloy. Widely used for stressed components in aircraft and elsewhere
—	0.5	—	—	0.6	—	Wrought Heat-treatable	Corrosion-resistant alloy for lightly stressed components such as glazing bars, window sections and automotive body components
1.8	2.5	1.0	—	0.2	0.15 Titanium 1.2 nickel	Cast Heat-treatable	Suitable for sand and gravity die-casting. High rigidity with moderate strength and shock resistance. A general purpose alloy
—	—	—	—	10.5	0.2 Titanium	Cast Heat-treatable	A strong, ductile and highly corrosion-resistant alloy used for aircraft and marine castings both large and small

and limitations of hot-working, cold-working and casting processes respectively. Generally speaking, cold-working refers to processes carried out at room temperature, that is, below the temperature of re-crystallisation for the metal or alloy which is being worked to shape. Hot-working implies that the metal is being formed at very much higher temperatures, often at red-heat, that is, above the temperature of re-crystallisation. *Recrystallisation* is considered in section 1.14, since an understanding of this phenomenon is required when considering the heat treatment processes called *annealing*.

Table 1.14 Soft solders

BS solder	Composition (%)			Melting range (°C)	Remarks
	Tin	Lead	Antimony		
A	65	34.4	0.6	183–185	Free running solder ideal for soldering electronic and instrument assemblies. Commonly referred to as electrician's solder
K	60	39.5	0.5	183–188	Used for high-class tinsmith's work, and is known as tinman's solder
F	50	49.5	0.5	183–212	Used for general soldering work in copper-smithing and sheet metal work
G	40	59.6	0.4	183–234	Blow-pipe solder. This is supplied in strip form with a D cross-section 0.3 mm wide
J	30	69.7	0.3	183–255	Plumber's solder. Because of its wide melting range this solder becomes 'pasty' and can be moulded and wiped

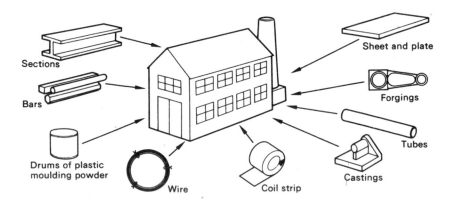

Fig. 1.18 Forms of supply

Table 1.15 Hot-working processes

Advantages	Limitations
1. Low cost	1. Poor surface finish — rough and scaly
2. Grain refinement from cast structure	2. Due to shrinkage on cooling the dimensional accuracy of hot-worked components is of a low order
3. Materials are left in the fully annealed condition and are suitable for cold-working (heading, benching, etc.)	3. Due to distortion on cooling and to the processes involved, hot-working generally leads to geometrical inaccuracy
4. Scale gives some protection against corrosion during storage	4. Fully annealed condition of the material coupled with a relatively coarse grain leads to a poor finish when machined
5. Availability as sections (girders) and forgings as well as the more usual bars, rods, sheets and strip and butt welded tube	5. Low strength and rigidity for metal considered
	6. Damage to tooling from abrasive scale on metal surface

Table 1.16 Cold-working processes

Advantages	Limitations
1. Good surface finish	1. Higher cost than for the hot-worked materials. It is only a finishing process for material previously hot-worked. Therefore, the processing cost is added to the hot-worked cost.
2. Relatively high dimensional accuracy	2. Materials lack ductility due to work hardening and are less suitable for bending, etc.
3. Relatively high geometrical accuracy	3. Clean surface is easily corroded
4. Work hardening caused during the cold-working processes: (a) increases strength and rigidity; (b) improves the machining characteristics of the metal so that a good finish is more easily achieved	4. Availability limited to rods and bars also sheets and strip, solid drawn tubes

Table 1.17 Casting processes (gravity, sand only)

Advantages	Limitations
1. Virtually no limit to the shape and complication of the component to be cast	1. Strength and ductility low, as structure is un-refined
2. Virtually no limit to the size of the casting	2. Quality is uncertain, as local differences of structure and mechanical defects such as blow-holes cannot be controlled or corrected
3. Low cost, as no expensive machines and tools are required as in forging	3. Low accuracy due to shrinkage
4. Scrap metal can be reclaimed in the melting furnace. (Wrought and machined components have to be made from relatively expensive pre-processed materials)	4. Poor surface finish
	5. Component must be designed without sudden changes of section, so that molten metal flows easily and cooling cracks and warping will not occur
	6. Not all metals are suitable for casting. The best metals have a low shrinkage, a short freezing range and high fusibility (melt at relatively low temperatures), and have a high fluidity when molten

1.12 Heat treatment of metals

From time to time in this chapter terms such as 'annealing', 'quench hardening', etc., have been used. These refer to the many *heat treatment* processes which can be used to modify the properties of a metal, that is, make it harder or softer; tougher or more brittle; stronger or weaker; more ductile or less ductile; as required. Figure 1.19 shows the procedures for heat treating plain carbon steels. In each instance the steel is heated above a critical temperature which is dependent upon the carbon content of the steel and the process. If this temperature is not achieved, the desired change in properties will not occur. If this temperature is exceeded, or held for too long a period of time, grain growth will occur and the steel will be weakened. This critical heating is followed by cooling. It can be seen from the diagram that the cooling rate is equally critical and that it alone determines the final properties of the heat treated metal.

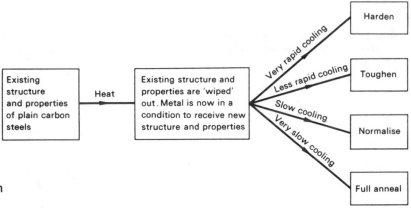

Fig. 1.19 Heat treatment of plain carbon steels

1.13 Hardening plain carbon steels

The above comments apply mainly to ferrous metals. The heat treatment of non-ferrous metals and their alloys varies according to the metal or alloy being treated (see section 1.17).

Figure 1.20 relates temperature and the carbon content to the structure and composition of steel. This diagram provides a basis for deriving the temperatures for heat treating plain carbon steels.

Ferrite is soft and ductile
Pearlite is tough
Cementite is hard and brittle

To quench harden a steel, it must be *quenched* (cooled very quickly) from the temperatures shown in Fig. 1.21 depending upon its carbon content. The degree of hardness achieved will be solely dependent upon:

(*a*) The carbon content.
(*b*) The rate of cooling.

The rapid cooling necessary to harden steel is known as *quenching*. The liquid into which the steel is dipped to cause this rapid cooling is called the *quenching bath*. In the workshop, the quenching bath

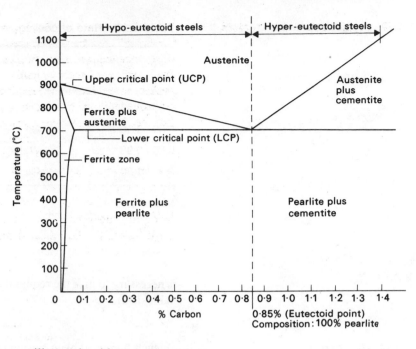

Fig. 1.20 Iron/carbon equilibrium diagram (steel section)

will contain either:

(i) water;

(ii) quenching oil (on no account use lubricating oil).

The more rapidly a plain carbon steel is cooled the harder it becomes for its partiuclar carbon content until the limit of the critical cooling rate is attained, after which there is no further increase in hardness. Unfortunately very rapid cooling also leads to *cracking* and *distortion*. Therefore the workpiece should not be cooled more rapidly than

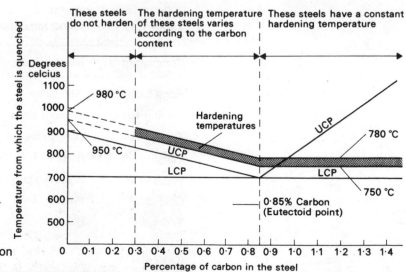

Fig. 1.21 Hardening of plain carbon steels

Table 1.18 Rate of cooling

Carbon content (%)	Quenching bath	Required treatment
0.30—0.50	Oil	Toughening
0.50—0.90	Oil	Toughening
0.50—0.90	Water	Hardening
0.90—1.30	Oil	Hardening

Note:
1. Below 0.5 per cent carbon content, steels are not hardened as cutting tools, so water hardening has not been included.
2. Above 0.9 per cent carbon content, any attempt to harden the steel in water could lead to cracking.

is necessary to give the required degree of hardness or toughness (see Table 1.18).

1.14 Tempering

Quench-hardened plain carbon steel is very brittle and unsuitable for immediate use. A further process known as *tempering* must be carried out to increase greatly the toughness of the steel at the expense of the loss of some hardness. Tempering consists of reheating the steel to a suitable temperature and again quenching it in water or oil. The temperature to which the steel is reheated depends upon the use to which the component is going to be put. Table 1.19 gives some suitable tempering temperatures for a selection of components made from plain carbon steel. The temper colour is the colour of the oxide film which appears on a freshly polished surface of the steel when it is reheated to the tempering temperature.

Table 1.19 Tempering temperatures

Component	Temper colour	Temperature (°C)
Edge tools	Pale straw	220
Turning tools	Medium straw	230
Twist drills	Dark straw	240
Taps	Brown	250
Press tools	Brownish-purple	260
Cold chisels	Purple	280
Springs	Blue	300
Toughening (crankshafts)	—	450—600

1.15   Recrystallisation

The terms hot- and cold-working have already been introduced in section 1.11 of this chapter. The term cold-working is relative. For instance, the metal lead hot-works at room temperature, whilst mild steel cold-works at temperatures up to 600 °C. For the majority of metals and alloys it can be assumed that they cold-work at room temperature. More precisely, the temperature at which a metal ceases to be cold-worked and becomes hot-worked is the temperature at which the crystals distorted by the forming process start to change back into normal crystals. This temperatures depends upon two factors:

(*a*) the metal or alloy under consideration;
(*b*) the degree of cold-working to which the metal or alloy has been subjected prior to recrystallisation.

The way in which recrystallisation occurs is shown, simply, in Figure 1.22. The cold-worked and distorted crystal contains stress points

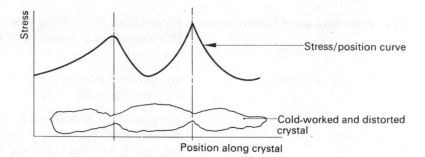

A cold-worked crystal is not of uniform cross-section. Therefore it will be subject to stress 'peaks' at the points of minimum cross-section

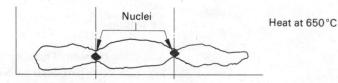

*At 650°C the nucleus of an equi-axed crystal will form at each stress 'peak'. Since 650°C is below the lower critical point this process is referred to as **sub-critical annealing***

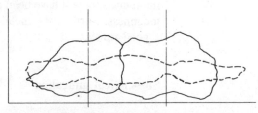

If heating is continued the nuclei grow – feeding on the material of the original cold worked crystal until it is consumed
The structure of the component will now consist of equi-axed crystals and it will be annealed
Note :   Since two or more new crystals replace each original crystal, they will be smaller. Grain refinement will have occurred

Fig. 1.22   Recrystallisation

locked into it. At a critical temperature, depending upon the metal and the degree to which the crystal has been distorted by cold-work, *nucleation* commences. That is, the nucleus of a new crystal starts to grow at each stress point. If the temperature of the metal is held at this temperature long enough, the new crystals will feed on the old crystals and grow until none of the distorted crystal structure is left and the metal consists entirely of new, undistorted crystals. This is called *recrystallisation*. Since there are several stress points in each distorted crystal, there are a larger number of smaller crystals after recrystallisation than at the start. Thus grain refinement also takes place.

Therefore, *cold-working* is the flow forming of metal *below* the temperature of recrystallisation for that metal. Similarly, *hot-working* is the flow forming of metal *above* its temperature of recrystallisation.

1.16 Annealing plain carbon steels

The annealing processes are used to soften steels that are already hard. This hardness may, in the case of ferrous metals, have been imparted in two ways:

(*a*) *Quench-hardening.*
(*b*) *Work-hardening.*

Work-hardening occurs when a metal is cold-worked. It becomes hard and brittle at the point where the cold-working is causing the crystal structure of the metal to become distorted. For example, if a strip of metal is bent backwards and forwards in a vice it starts to harden at the point of bending and eventually breaks off.

Annealing

Annealing consists of heating the metal to the temperatures shown in Fig. 1.23 for its particular carbon content, followed by very slow cooling. Usually this is done by turning the furnace off with the work in it, closing the dampers, and allowing work and furnace to cool down together. Obviously there will be appreciable grain growth, and maximum ductility will have been achieved at the expense of strength and toughness. Further, the metal will tend to be too soft to machine to a good finish.

Sub-critical annealing

Sub-critical annealing consists of heating the work-hardened steel components until their temperature lies within the 630° to 700 °C band as shown on Fig. 1.23. The components are then soaked until the required degree of recrystallisation and grain growth has been achieved,

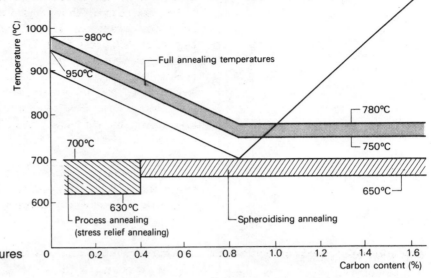

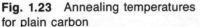

Fig. 1.23 Annealing temperatures for plain carbon

after which they are then cooled slowly. This process cannot be used to anneal quench-hardened components, as such components will not have received the work-hardened, distorted, stressed crystals necessary to trigger off the recrystallisation process. Although full annealing could be used to soften work-hardened components, sub-critical annealing is much more efficient. Since the temperatures used are much lower, furnace maintenance and fuel costs are much reduced. Further, little oxidation of the work occurs at these relatively low temperatures.

Spheroidising annealing

This is used for steels above 0.4% carbon. Such steels are unlikely to have been substantially cold-worked because of their relative lack of ductility. However they are likely to have been quench-hardened. Thus they are not in a suitable condition for sub-critical annealing. Where a quench-hardened steel component requires to be softened before re-machining, spheroidising annealing is to be preferred to full annealing. Not only is it carried out at reduced temperatures with the benefits listed under sub-critical annealing, but also the finer grain structure gives better machining properties. The temperature range for this process is shown in Fig. 1.23. The components are soaked at this temperature until spheroidisation is complete and then cooled slowly. They will have a fine grain structure which will machine to a good finish, but they will lack the ductility for cold-working. If even limited cold-working is required then full annealing must be resorted to.

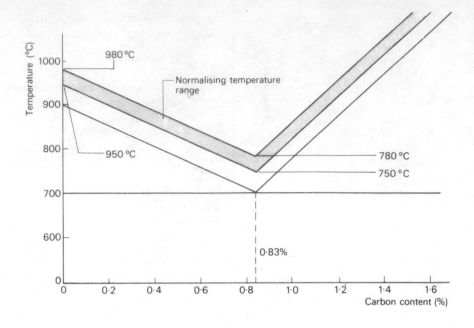

Fig. 1.24 Normalising temperatures

1.17 Normalising

The temperatures associated with the *normalising* of plain carbon steels are shown in Fig. 1.24. It can be seen that they differ from those for annealing in the hyper-eutectoid range. This renders the steel fully austenitic and free from stress. Cooling is quicker than for annealing, with the work being removed from the furnace followed by cooling in free air away from draughts. This faster rate of cooling results in a finer grain structure suitable for machining but lacking the ductility for extensive cold-working. Normalising is frequently used for stress relieving large components (particularly forgings) between rough machining and finish machining to remove any residual stresses which otherwise might cause distortion and dimensional instability in the finished component.

1.18 Heat treatment of non-ferrous metals and alloys

None of the non-ferrous metals and only a few non-ferrous alloys can be quench-hardened like plain carbon steel. The majority of non-ferrous components are hardened by manufacturing them by a cold-working process, or manufacturing them from cold-rolled ('spring temper') sheet or cold-drawn wire and rod. Some alloys, notably those based upon aluminium, age-harden naturally after annealing or they can be hardened more quickly by a precipitation heat treatment process. Since, in the main, most hardened non-ferrous metals are work-hardened it follows that they must be annealed by a recrystallisation process similar to the sub-critical annealing process for steel. However, non-ferrous metals and alloys do not have to be cooled slowly after recrystallisation and, in fact, there is some advantage in quenching them. The sudden cooling and resulting shrinkage strips the oxide film from the surface of the metal leaving it clean. Also the rapid cooling improves the mechanical properties as it prevents excessive grain growth. Annealing temperatures for typical non-ferrous materials which have been hardened by cold-working are as follows:

Aluminium   500−550 °C   (pure metal)
Copper   650−750 °C   (pure metal)
Cold-working brass   600−650 °C   (simple alloy)

Those non-ferrous alloys which can be hardened by natural ageing and/or precipitation treatment require a different approach. They are softened by *solution* treatment. Typical of this group of non-ferrous alloys is *duralumin*. This is an aluminium based alloy containing copper, magnesium, manganese and zinc, and which is suitable for cold-working (a 'wrought' alloy, as opposed to a 'casting' alloy).

Solution treatment

The duralumin alloy is raised to a temperature at which the aluminium content can form solid solutions with the alloying elements, hence the name of the process. The solution treatment temperature is approximately 500 °C after which the alloy is quenched to preserve the solid solution. Solid solutions are soft and ductile, and after solution treatment the metal is in a suitable condition for cold working.

Precipitation treatment

If the alloy is kept at room temperature after solution treatment, hardening commences spontaneously. After twenty four hours there is a risk of cracking and after four days full hardness is achieved. This is caused by particles of a copper-aluminium inter-metallic compound precipitating out of the solid solution. Since inter-metallic compounds are hard and brittle, their presence renders the metal hard and brittle, lacking in ductility, and unsuitable for cold working.

When hardening occurs naturally it is referred to as *age-hardening* and, as has been stated, maximum hardness occurs after four days. Natural ageing can be delayed by refrigerating the solution-treated components at −6 to −10 °C. Precipitation hardening can be accelerated by reheating the solution-treated components to about 150 to 170 °C for a few hours depending upon the alloy. This is referred to as *artificial ageing*. Both natural age hardening and artificial ageing are *precipitation hardening* processes. The times and temperatures for each alloy and the maker's specified treatment must be carefully observed.

1.19 Polymeric (plastic) materials

There is an ever increasing range of synthetic materials available under the general name of *plastics*. These materials rarely show plastic properties in their finished condition, in fact many of them are elastic, but during the moulding operation by which they are formed they are reduced to a plastic condition by heating to just above the temperature of boiling water. There are two main groups of plastic materials, thermosetting plastics and thermoplastics.

Thermosetting plastics

This group undergoes a chemical change called polymerisation during moulding and they can never again be softened by re-heating. These materials are hard, rigid and rather brittle. The polymerisation may be caused by heating the moulds during the forming operation, or by the addition of a suitable chemical called a hardener to the plastic material immediately prior to use. A typical 'plastic' moulding powder may consist of:

Plastic resin	38% by weight
Filler	58% by weight
Pigment	3% by weight
Mould release agent	0.5% by weight
Catalyst	0.3% by weight
Accelerator	0.2% by weight

The filler has a considerable influence on the properties of the mouldings produced from a given material. It improves the impact strength and reduces shrinkage during moulding. It also substantially reduces the overall material cost. Typical fillers are:

Glass fibre	high strength and good electrical insulation properties;
Wood flour	low cost, low strength, high bulk;
Calcium carbonate	low cost, low strength, high bulk;
Shredded paper	low cost, medium strength, high bulk, fair electrical properties;
Shredded cloth	low cost, high strength, high bulk, fair electrical properties;
Mica granules	medium cost, high strength, high bulk, good heat resistance and electrical properties.

The pigment gives colour to the finished properties. It can also filter out the ultra-violet rays of sunlight and delay degradation. The mould release agent prevents the moulding from sticking to the mould. It also acts as an internal lubricant and helps the moulding powder to flow into the finer recesses of the mould during the moulding process. The catalyst promotes the curing process during moulding and ensures uniformity of the properties. The accelerator speeds up the curing process and shortens the time of the moulding cycle and increases productivity. It may also include a stabiliser which prevents curing at room temperature whilst not affecting it at moulding

Table 1.20   Some typical thermosetting plastic materials

Material	Characteristics
Phenolic resins and powders	These are used for dark-coloured parts because the basic resin tends to become discoloured. These are heat-curing materials
Amino (containing nitrogen) resins and powders	These are colourless and can be coloured if required; they can be strengthened by using paper-pulp fillers, and used in thin sections
Polyester resins	Polyester chains can be cross-linked by using a monomer such as styrene; these resins are used in the production of glass-fibre laminates
Epoxy resins	These are also used in the production of glass-fibre laminates

temperatures. Some typical thermosetting plastics are listed in Table 1.20.

Thermoplastics

These can be softened as often as they are re-heated. However some degradation occurs if they are over-heated or heated too often and recycled material is usually only used for lightly stressed components. They are not as rigid as thermosetting plastics but are generally tougher. Some typical thermoplastics are listed in Table 1.21.

Laminated plastic (Tufnol)

Fibrous materials such as paper or woven cloth are impregnated with a thermosetting plastic resin. The impregnated sheets are then layed up in a hydraulic press between highly polished metal sheets until the desired thickness is obtained. The direction of the grain is rotated through ninety degrees for each laminate so that the properties are the same in all directions. When the required number of laminations has been layed up the press is closed and the resin cures into a homogeneous sheet of reinforced plastic material of great strength. As well as sheet, this material can also be supplied as rods, tubes, and simple mouldings. This material can be machined dry with

Table 1.21 Some typical thermoplastic materials

Type	Material	Characteristics
Cellulose plastics	Nitrocellulose	Materials of the 'celluloid' type are tough and water resistant. They are available in all forms except moulding powders. They cannot be moulded because of their flammability
	Cellulose acetate	This is much less inflammable than the above. It is good for tool handles and electrical goods
Vinyl plastics	Polythene	This is a simple material that is weak, easy to mould, and has good electrical properties. It is used for insulation and for packaging
	Polypropylene	This is rather more complicated than polythene and has better strength
	Polystyrene	Polystyrene is cheap, and can be easily moulded. It has a good strength but it is rigid and brittle and crazes and yellows with age
	Polyvinyl chloride (PVC)	This is tough, rubbery, and practically non-inflammable. It is cheap and can be easily manipulated: it has good electrical properties
Acrylics (made from an acrylic acid)	Polymethyl methacrylate	Materials of the 'perspex' type have excellent light transmission, are tough and non-splintering, and can be easily bent and shaped
Polyamides (short carbon chains that are connected by amide groups-NHCO)	Nylon	This is used as a fibre or as a wax-like moulding material. It is fluid at moulding temperature, tough, and has a low coefficient of friction
Fluorine plastics	Polytetrafluoroethylene (ptfe)	Is a wax-like moulding material; it has an extremely low coefficient of friction. It is very expensive
Polyesters (when an alcohol combines with an acid, an 'ester' is produced)	Polyethylene terephthalate	This is available as a film or as 'Terylene'. The film is an excellent electrical insulator

ordinary engineering cutting tools using low rake and a high cutting speed. It is widely used for bearings, gears and other engineering components which have to operate in hostile environments. It is also used for heavy duty insulators and, when coated with copper, is used as the basic material for printed circuit boards.

Glass reinforced plastic (GRP)

Woven and chopped strand glass fibre can be bonded together by epoxy and polyester resins to form large and complex mouldings from crash helmets to 18-metre yachts and fishing vessels. The resin used to bond the fibres together is a thermosetting plastic and is set by chemical action at room temperature. No press is required. The impregnated glass fibre is laid up over wooden or plastic patterns or in fibre glass or plaster moulds. When set the moulding is lifted off and the pattern or mould is used again. As well as being widely used for small boat building it is also used for specialist car bodies, large casting patterns, copy milling models, machine guards and high quality printed circuit boards.

1.20 Properties of plastics

The properties of plastics can vary widely in respect of strength, toughness, softening temperature, etc., but all plastic materials have the following properties in common.

Electrical insulation

All plastic materials exhibit good electrical insulation properties. Unfortunately, their usefulness in this field is limited by their low heat-resistance and softness. They are therefore useless as formers to wind electric radiator elements and as insulators for outdoor transmission lines as their surface would soon be roughened by the weather. Dirt collecting on this roughened surface would then provide a conductive path, causing a short circuit. However as insulators for flexible and semi-rigid cables and in particular for cables carrying telecommunication signals, plastic materials are excellent. They can also be easily moulded into switch gear components and housings for electrical and electronic components and equipment.

Strength/weight ratio

Plastic materials vary in strength considerably. Some of the stronger, such as nylon, compare favourably with the weaker metals. However because of their lower density, properly designed and proportioned components can be as strong as most metals on a strength/weight ratio.

Thus they are steadily taking over engineering duties which, until recently, were considered the prerogative of metals.

Corrosion resistance

Plastic materials are inert to most inorganic chemicals. Thus they can be used in environments which are hostile to the most corrosion resistant metals. They are also superior to natural rubber as they are resistant to attack by oils and greases.

It should be noted that because of the ease with which plastics can be manufactured and used, they are finding increasing use as pipes and conduits. Care must be taken when inserting a length of plastic pipe or conduit into any system, in place of metal, that the earth continuity of the system is not destroyed. This can occur because cold water pipes and conduits are often used as convenient earthing points for electrical equipment.

1.21 Bearing materials (anti-friction)

There are essentially two types of bearings. Those where the moving parts slide over each other as in plain bearings, and machine tool slideways, and those where the moving parts roll over each other as in ball and roller bearings. Metals and non-metals developed as bearing materials for the former type (sliding contact) are listed in Table 1.22, and require the following properties. Note that with sliding bearings there should always be a lubricant present to prevent actual contact between the bearing surfaces.

(a) Coefficient of friction (μ)

This should be kept as low as possible to avoid wasting energy. A sledge runs down hill much more easily on snow than on the grass which is exposed when the snow melts. This is because the snow (ice) has a very low coefficient of friction. Energy wasted in overcoming friction causes the bearing to heat up. Eventually the bearing may overheat and be destroyed.

(b) Strength

The bearing material must have sufficient strength to support the shaft and any load that may be applied to the shaft in service.

(c) Wear resistance

The bearing material must resist wear to reduce maintenance. However it is better for the bearing to wear out before the journal on the shaft,

Table 1.22 Some typical bearing materials (sliding contact)

Category	Sn	Sb	Cu	Pb	P	Properties and applications
White metal	93	3.5	3.5	—	—	Big-end bearings for light and medium duty, high-speed internal combusion engines
	86	10.5	3.5	—	—	Main bearings for light and medium duty, high-speed internal combusion engines
	80	11.0	3.0	6.0	—	General purpose, heavy duty bearings. Lead improved plasticity where alignment is a problem
	60	10.0	28.5	1.5	—	Heavy duty marine reciprocating engines, electrical machines
	40	10.0	1.5	48.5	—	Low cost, general purpose, medium duty, bearing alloy
Bronze	10.5	—	89	—	0.5	Good anti-friction properties, suitable for heavy loads, rigid
	10.0	—	79.9	10	0.1	Good anti-friction properties, lubrication not critical, lead content reduces rigidity and helps alignment
	3	—	74	23	—	Leaded (plastic) bronze, excellent self-alignment properties due to high lead content. For duty intermediate between white metal and phosphor bronze

Category	Fe	C	Si	Mn	S:/P	Properties and applications
Cast iron	94	3.3	1.3	1.0	0.1/0.3	The flakes of graphite (carbon) in grey cast iron gives it self lubricating properties. Suitable for heavy duty, low-speed applications where lubrication is difficult, e.g. machine tool slideways

Category	Composition	Properties and applications
Plastic	Polytetrafluorethylene	Teflon: Can withstand much higher temperatures than most plastics. Very expensive anti-friction coating — very low coefficient of friction. Does not require lubrication
	Polyamide	Nylon: Can be moulded into bushes and gears. Does not require lubrication. Use for office and food processing machinery
	High density polyethylene	Low cost bearings. Does not require lubrication. Cannot support such high loads as Nylon or Teflon

Sn = Tin, Sb = Antimony, Cu = Copper, Pb = Lead, P = Phosphorus, Fe = Iron, C = Carbon, Si = Silicon, Mn = Manganese, S = Sulphur

since it is cheaper and easier to replace the bearing shell than to replace the shaft.

(d) Plasticity

It is virtually impossible to obtain perfect alignment between a shaft and its associated bearings. Therefore bearing materials should be capable of slightly distorting and bedding in so that optimum alignment is ensured early in the life of the assembly. White metals and leaded bronzes are better at aligning themselves than the harder and more rigid phosphor bronzes.

(e) Surface texture

A perfectly smooth surface would be unsuitable for a bearing as there would be no provision for the retention of pockets of lubricant. An ideal bearing material should have a 'textured' surface consisting of hard facets of anti-friction material dispersed through a soft matrix. The matrix wears away between the facets to form pockets for the retention of lubricant, whilst the shaft is supported by the hard, wear resistant facets of anti-friction material. The soft matrix also has sufficient 'give' to assist alignment.

(f) Corrosion resistance

The bearing material should resist corrosion by impurities in the lubricant or impurities which may enter the bearing in service. It should also resist attack from any additives (sulphur or chlorine-based extreme pressure additives) in the lubricant intended to give it greater lubricity. Corrosion eats into the bearing material changing its texture and reducing its efficiency and, in extreme cases, the bearing material may be destroyed completely.

(g) Thermal conductivity

Since even the best of bearings offer some friction, there will always be some energy loss and a corresponding rise in temperature when the shaft is rotating. This heat energy can only escape through the lubricant or by conduction through the walls of the bearing. If the heat is not conducted away quickly enough, the temperature will rise and the bearing material may melt. Thus it is important that the bearing is designed for easy dissipation of any heat energy generated within it. Plastic bearing materials are at a disadvantage in this respect because of their poor thermal conductivity.

1.22 Rubbers

Rubbers (elastomers) are substances which permit extreme reversible extensions to take place at normal temperatures. Natural rubber (isoprene) is an obvious and important elastomer. Although the rubbers do not possess the strength of many other polymer (plastic) materials, they are widely used for applications where softness, flexibility and elongation are of primary importance together with a high coefficient of friction and resistance to environmental degradation. Some typical applications are: drive belts, floor coverings, oil seals, gaskets, vehicle tyres, and anti-vibration machine mountings. Most synthetic rubbers become brittle below -20 °C, with the exception of silicone rubber which becomes brittle at -60 °C. Natural rubber also becomes brittle at -60 °C. Because it is readily attacked by solvents, oils, ozone, and petrol, and because it degrades (perishes) in the presence of sunlight, natural rubber is not much used by engineers. The synthetic rubbers such as neoprene, are widely used in engineering since they have good resistance to oxidation, ageing and weathering. Neoprene, in particular, is resistant to oils and solvents, abrasion and elevated temperatures. Because of its chlorine content it is fire resistant. It is used as a flexible electrical insulator, and for gaskets, hoses, engine mountings, sealants, rubber cements and protective clothing.

The uses to which rubbers (elastomers) may be put in engineering may be classified as follows:

Vibration insulation and isolation
(a) Shock absorbers.
(b) Anti-vibration machine mounts.
(c) Sound insulation.

Distortional systems
(a) Correctives for misalignment such as flexible couplings for shafts.
(b) Changing shapes such as belts, flexible hose, covered rollers, tyres, etc.
(c) Seals of all kinds, and gaskets.
(d) Rubber hydraulics (rubber form tools for sheet metal).

Protective systems
(a) Protection against abrasion.
(b) Protection against corrosion.
(c) Electrical insulation.

1.23 Ceramics

'Ceramics' is the name given to a wide range of inorganic materials which include glass, fire bricks for furnace linings, porcelain high-voltage insulators, abrasive grits for grinding wheels and cutting tool tips. There are four main groups of ceramic materials.

Amorphous ceramics

These are substances generally referred to as *glass*.

Bonded ceramics

These are materials in which hard crystals are bonded together in a glass-type matrix. These materials are used for cutting tool tips.

Cements

These may be crystalline or a combination of crystalline and amorphous materials. They are used as bonding materials in the construction industry and as 'fire-clays' for fixing fire bricks in a furnace lining.

Crystalline ceramics

These are materials such as magnesium oxide which is used as an electrical insulator in copper sheathed mineral insulated cables capable of operating at high temperatures, and aluminium oxide (emery) which is used as an abrasive. The hard carbides and nitrides used in cutting tool tips are also classified as crystalline ceramics, but do not exist naturally.

Properties

Ceramic materials are very hard compared with other engineering materials which is why they are of such interest as cutting tool materials. Although they have a high compressive strength they lack ductility and tensile strength and are brittle. Thus great care must be taken in supporting cutting tool tips made of these materials so that all the cutting forces are compressive. Care must also be taken in presenting the tool to the work so that the cutting edge is not chipped. Thus such tools are more successful when used on automatic machines where tool approach can be accurately controlled, than on manually controlled machines. The hardness of ceramics can also be exploited for high-voltage electrical insulators used out of doors. Their hard and highly glazed surfaces do not become roughened by weathering and therefore remain clean and free from deposits of dirt which could cause conductive paths over the surface of the insulators.

Ceramics are also capable of withstanding high temperatures and are, therefore, good refractory materials. The most widely used materials are clays containing silicon oxide and aluminium oxide. Although they do not have the very high melting points of some of the more specialised ceramics, their availability and low cost make them suitable where large quantities of the material are required as in furnace linings. Unfortunately fire bricks made from these materials soften gradually over a range of temperatures and may collapse well below their normal melting point.

The glasses are a widely used group of ceramics. They are used for optical purposes such as lenses and fibre optics, as woven fibre or chopped strand mat for reinforcing plastics (GRP), and for laboratory apparatus such as test tubes and beakers, etc.

1.24 The corrosion of metals

Corrosion is the slow but continuous eating away of metallic components by chemical or electro-chemical attack. There are three factors which govern the rate of corrosion:

(a) the metal from which the component is made,
(b) the treatment which the surface of the component receives,
(c) the environment in which the component is kept.

All metals corrode to a greater or lesser extent; even precious metals like gold and silver tarnish in time and this is a form of corrosion. Corrosion-prevention processes are not able to eliminate the inevitable failure of a component by corrosion but, before this occurs, the treatment can have slowed down the corrosion process to a point where the component will have worn out or been discarded for other reasons.

Most of the cost of corrosion and its prevention is related to atmospheric corrosion. The corrosion problems related with chemical engineering, marine engineering and food processing are highly specialised and the remedies are more specific. In any case, the cost, although high, is negligible compared with the cost of atmospheric corrosion. This chapter will only deal with atmospheric corrosion.

Any metal exposed to normal atmospheric conditions becomes covered with a thin film of moisture. This moisture film is invariably contaminated with dissolved solids and gases which are ever present in the atmosphere and which increase the rate of corrosion. Hence the rate of corrosion depends to a large extent where it is taking place. Table 1.23 shows the rates of corrosion for unprotected steelwork in various environments.

The most common example of corrosion due to moisture and atmospheric oxygen is the rapid surface formation of *red rust* on iron

Table 1.23 Rate of corrosion

Type of environment	Typical rate of rusting for mild steel in temperate climates (mm/year)
Rural	0.025–0.050
Urban	0.050–0.100
Industrial	0.100–0.200
Chemical	0.200–0.375
Marine	0.025–0.150

and steel. This 'red rust' is an oxide of iron, but of different composition to the blue-black oxide of iron called 'mill scale' and formed when iron and steel is heated to high temperatures in a dry atmosphere. The conditions for rusting at normal room temperatures are a combination of *air and moisture*. Rusting will not take place in dry air, nor will it take place in water free of dissolved air (boiled water). Once 'rusting' commences the action is self generating. Removing contact with the original supply of water and air by painting over the rust will not stop this corrosion. The rusting will still continue under the paint. This is why surface treatment is so important and why all traces of rust must be removed before painting. The rate of rusting slows as the layer of rust deepens or thickens, but as rain washes off the surface the rust rate increases again. This cycle is continuous and, once started, it is very difficult to control.

Atmospheric pollution rapidly increases the rate of rusting of iron and steel. It also attacks, but more slowly, copper and zinc. Lead is virtually unaffected and so is aluminium if its surface has been correctly pre-treated and it is regularly washed clean. Figure 1.25 shows how atmospheric pollution is caused by the burning of fossil fuels. Other pollutants from chemical works and from sea-water spray in coastal districts aggravate the problem.

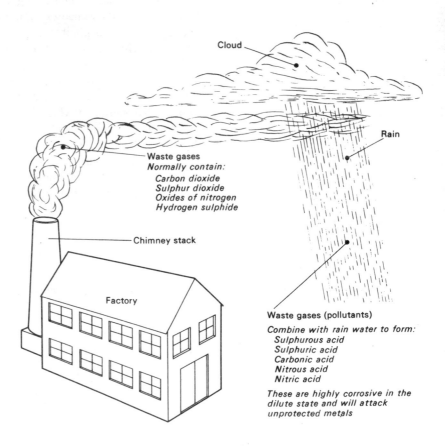

Fig. 1.25 Corrosive pollution

1.25 Corrosion prevention

Methods of preventing or retarding corrosion are mostly applied to iron and steel since most of the non-ferrous metals and alloys form their own protective coatings as previously described.

For short-term protection iron and steel components may be coated with oil or grease. However, in the long-term this is not satisfactory for the following reasons:

(a) the protective film will dry up and no longer seal the surface from atmospheric attack;

(b) oils and greases gradually absorb moisture from the atmosphere and corrosion can take place *under* the oil or grease film;

(c) cheap oils and greases often contain active sulphur and acid impurities that will themselves attack and corrode the very metal surface which they are supposed to be protecting.

Care must always be taken when two dissimilar metals come into contact with each other in a moist or wet environment especially if polluted. Under these conditions they behave like a simple electric battery and the currents generated cause corrosion. Metals can be arranged in a special order called the *electro-chemical series*. This series is given in Table 1.24 and, it should be noted, that in this context hydrogen gas behaves like a metal. If any two metals from the table come into contact in a moist and polluted environment, the more negative one will corrode the most rapidly. For example:

(a) in galvanised iron (zinc coated steel) the zinc is more negative than the iron in the steel so the zinc corrodes away whilst protecting the steel. The zinc is said to be *sacrificial*;

Table 1.24 Electro-chemical series

Metal	Electrode potential (volts)	
Sodium	−2.71	*Corroded (anodic)*
Magnesium	−2.40	
Aluminium	−1.70	
Zinc	−0.76	
Chromium	−0.56	
Iron	−0.44	
Cadmium	−0.40	
Nickel	−0.23	
Tin	−0.14	
Lead	−0.12	
Hydrogen (reference potential)	0.00	
Copper	+0.35	
Silver	+0.80	
Platinum	+1.20	
Gold	+1.50	*Protected (cathodic)*

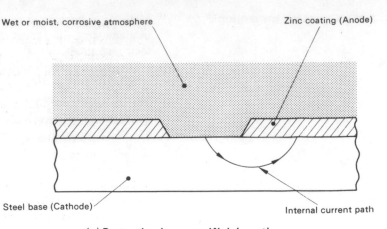

(a) **Protection by a sacrificial coating**

Coating is eaten away whilst protecting the base

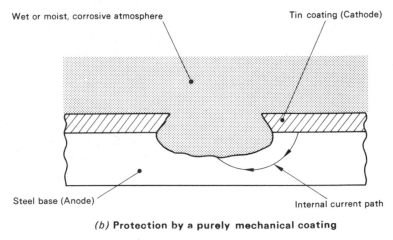

(b) **Protection by a purely mechanical coating**

Coating only protects the base if intact.
If coating is damaged, base is eaten away quicker than if coating were not present.

Fig. 1.26   Electrolytic corrosion

(*b*) in tin-plate, the mild steel is corroded if the tin coating is broken at any point, since the iron in the steel is more negative than the tin. Hence the cut edges of tin-plate should be sealed with solder or painted with a lacquer, and marking out should be done with a pencil and not with a scriber. Figure 1.26 shows what happens during electrolytic corrosion.

Chemical and electro-chemical corrosion is intensified when a metal is under stress. The stresses involved are usually internal stresses caused by some cold forming process. The corrosive attack is usually along the crystal boundaries and this weakens the metal far more than surface corrosion. Stress-relief by heat treatment after manufacture prevents this form of corrosion. The following points should be observed during the design stage of a component or assembly to reduce corrosion to a minimum.

(a) Prevention of crevices or moisture traps.

(b) Selection of a suitable material which is inherently corrosion resistant, or the close specification of an anti-corrosion treatment process for the material chosen.

(c) Sealing joints if they are not continuously welded.

(d) Adequate ventilation and drainage.

(e) Contact with corrosive substances to be kept to a minimum.

(f) Ease of washing down and cleaning.

1.26 Surface preparation

It has been firmly established that the essential and most important factor of any efficient anti-corrosion treatment is surface preparation prior to the treatment. However carefully selected the protective process may be, it cannot fulfil its purpose if it is applied to an inadequately prepared surface. Surfaces carrying dirt, grease, corrosion or mill-scale are unsuitable for direct application of an anti-corrosion treatment. Table 1.25 describes how the surface may be prepared for treatment.

1.27 Galvanising

This is the coating of mild steel with zinc. There are two alternative processes: *hot dip galvanising*, in which the cleaned and fluxed work is dipped into a bath of molten zinc; or *electrolytic galvanising*, where the zinc is deposited elctrolytically on the sheet metal base.

Hot dip galvanising is a very versatile process. It can be applied equally well to structural steelwork, nuts and bolts, strip, tube, wire and cast iron. It is widely used because of its reliability, its ability to withstand rough treatment, its relatively low cost, and the unique type of corrosion resistance (sacrificial coating) provided by all methods of zinc coating. Wherever possible *finished* fabricated components and assemblies should be hot dip galvanised to seal any raw edges and seal all joints. This treatment is excellent for utensils used out of doors — e.g. agricultural hardware. The work to be treated is prepared by pickling it in hot sulphuric acid or cold hydrochloric acid to clean the surface. It is then fluxed with a mixture of zinc chloride and ammonium chloride. Finally the work is dipped into a bath of molten zinc to which a trace of aluminium has been added to give the traditionally bright finish as well as ensuring a coating which is smooth and of uniform thickness. The work is then quenched in cold water to cleanse the surface and make the work cool enough to handle.

In electrolytic galvanising the zinc is deposited on the work electrically. This gives greater control over the thickness and uniformity of the coating. It is also cheaper since the energy required to deposit the zinc electrolytically is less than the energy to keep a large bath of zinc molten and make good the heat loss when cold work is dipped into it. The coating deposited by electrolytic galvanising is much thinner than for hot dip galvanising and, usually, it is used as a pre-treatment for painting or plastic coating.

Table 1.25   Surface preparation

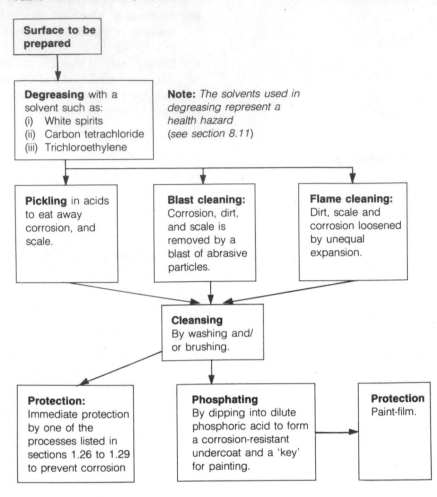

1.28 Electro-plating

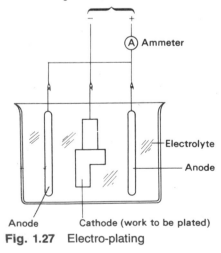

Fig. 1.27   Electro-plating

It has just been stated that in electrolytic galvanising the zinc is deposited on the work electrolytically. Many other metals apart from zinc can be plated onto components electrolytically (e.g. chromium, copper, nickel and tin) and this process is called electro-plating. It is used to cover a component with a coating which may be decorative or protective or both. The components to be plated are immersed in a solution called an electrolyte and are connected to the *negative* pole of a direct current (dc) supply. This makes them the *cathodes* of the plating cell. This is shown in Fig. 1.27. To complete the circuit, *anodes* connected to the *positive* pole of the direct current supply are also immersed in the electrolyte.

The electrolyte provides the metal ions which are deposited onto the surface of the work. The anodes may be *soluble*, in which case they are made of the same metal as that being deposited, and they dissolve into the electrolyte to replace the metal ions deposited on the work. Thus the strength and balance of the electrolyte is maintained. This applies in the cases of nickel, copper and zinc plating. Alternatively, the anodes may be *insoluble* as in chromium plating. In this case the anodes only complete the circuit and the electrolyte is steadily weakened.

The mass of metal deposited is proportional to:

(*a*) the magnitude of the current in amperes passing through the plating cell;
(*b*) the time in seconds for which the current flows;
(*c*) the electro-chemical equivalent of the metal.

1.29 Anodising

Aluminium relies upon an oxide film on the surface of the metal to resist corrosion. The process of anodising artificially builds up a thick, adherent layer of aluminium oxide which is resistant to atmospheric corrosion both for interior and exterior purposes even when subjected to the pollution of urban atmospheres.

Components to be anodised are first cleaned and degreased by the use of chemical solvents, after which they are etched or polished depending upon the surface texture required. The work is then made the *anode* of an electrolytic cell (the reverse of electro-plating) and a direct electric current is passed through the work. The electrolyte is a dilute acid and typical acids and the types of finish they provide are as follows.

Sulphuric acid

The oxide film produced by sulphuric acid is widely used for protection against general atmospheric and marine corrosion. The finish is colourless and the work is dyed after anodising if any particular colour is required.

Oxalic acid

This gives a film which is as corrosion resistant as that obtained with sulphuric acid, but which is harder and more wear resistant. Oxalic acid is largely used on the continent of Europe for anodising architectural metalwork. The colour obtained is the natural film colour (integral colour) and is not a dye.

Chromic acid

This was the first anodising process to be used and is still widely used for anodising aircraft components — particularly riveted assemblies

— where entrapped acid could cause corrosion. Chromic acid produces *minimum* corrosion under such conditions. The film produced provides an excellent 'key' for subsequent painting processes.

Selected sulphonated aromatic acids

These have been developed to provide a wide range of integral colours which are superior to dyed colours and highly resistant to weathering. After treatment the surface is finally sealed, usually with boiling water or steam, to improve the corrosion resistance of the coating and to minimise the absorption properties of the oxide film. Depending upon the process used, anodised surfaces can be produced with a wide range of physical properties and decorative appearances to satisfy any given set of service conditions.

1.30 The paint film

Painting is widely used for the protection and decoration of metallic components and structures. It is the easiest and cheapest coating which can be applied with any degree of permanence and, by careful choice, painting can provide a wide range of protective properties. Painting can also be used as a sealant over such finishes as galvanising and anodising. This is particularly useful in urban areas where the sulphur in the atmosphere rapidly destroys the sacrificial zinc coating of galvanised steelwork.

Paints may generally be described as consisting of finely divided solids (pigments) in a liquid suspension which dries or sets to provide a coherent film over the metal surface. Usually a paint is made up of three main constitutents.

The vehicle: This contains the film forming component or 'binder' in a volatile solvent. The binder is a natural or synthetic resinous material and reflects the essential properties of the paint: its durability, protective ability, flexibility and adhesion.

The pigment: This provides the paint with its opacity and colour. Further, some pigments have special properties and act as corrosion inhibitors, fungicides, insecticides, etc.

The solvent or thinner: This controls the consistency of the paint and controls its application. Since the solvent evaporates once the paint has been spread it forms no part of the final film. In addition, a paint may contain small quantities of a catalyst or accelerator to speed up the drying reactions, together with anti-skinning, anti-setting, and thixotropic (anti-drip) agents.

A complete paint system consists of the following.

(a) *A primer*, which is used as an adhesive for the subsequent protective coats. Since it may also contain a corrosion inhibiter, it should be matched to the material being painted. Primers used on some metals, such as aluminium, contain an etching agent to produce a suitable 'key' so that the primer will adhere to the metal sub-strata.

(b) *Putties or fillers*, which are applied with a knife or spatula to fill surface defects in castings, and dents and blemishes in sheet metal.

(c) *Undercoats*, which are used to build up the thickness of the paint film, cover the primer and filler and give opacity to the colour of the finish coats, as well as providing a smooth surface for their application where a high gloss is required (e.g. motor car bodywork).

(d) *Finish or top coats*, which are not only decorative but provide most of the corrosion resistance. This is because they contain a 'varnish' which seals the under coats and prevents the absorption of moisture. They are usually tough and abrasion resistant, being based on acrylic or polyurethane rubbers.

1.31 Types of paint

Paints can be broadly classified, by the manner in which they dry, into four groups.

Group 1

In this group, atmospheric oxygen reacts with the binder causing it to polymerise into a solid film. This reaction is speeded up by *forced drying* at 70 °C. Paints that dry by oxidation include the traditional linseed-oil based paints, the oleo-resinous paints, and the modern general purpose air drying paints based on oil-modified alkyd resins.

Group 2

These paints are based upon amino-alkyd resins which do not cure (set) at room temperature and have to be 'stoved' at 110−150 °C to promote the polymerisation reaction. When set, such paints are tougher and more resistant to abrasion than air drying paints. Paints in this group are used for motor car bodies.

Group 3

In this group, polymerisation is caused by the addition of an activator or hardener. Since this is stored separately and only added to the paint immediately before use, such paints are referred to as 'two pack' paints. Polymerisation (hardening) commences as soon as the hardener is added to the paint. At first this will be slow, but, as soon as the paint is spread, a solvent commences to evaporate increasing the concentration of the hardener. This increase in concentration results in rapid polymerisation and the paint is soon 'touch-dry'. However, it does not attain its full mechanical properties and resistance to damage until after a few days. Paints in this category are based upon polyester, polyurethane, and epoxy resins. The tendency, nowadays, is to use 'one-can' paints. The hardener is added at the time of manufacture, but at too low a concentration to cause the polymerisation reaction to commence. As in the previous example, once the paint is spread, a volatile solvent evaporates and increases the concentration of the hardener to the level at which hardening commences.

Group 4

These are the 'lacquers', that is, paints which dry by the simple evaporation of the volatile solvent (thinners), with no hardening or polymerisation reaction taking place. Suitable resins for lacquers are cellulose nitrate and acrylics.

1.32 Preparation for painting

As for any of the protective coatings described so far, the success of a paint film largely depends upon satisfactory preparation of the metal surface being treated. This may consist of simple purging of the surface by mechanical or chemical means to remove scale, rust, grease and dirt. It may also include more sophisticated pretreatment such as galvanising or phosphate coating of the steelwork to be painted. The purpose of the surface preparation is not only to provide a 'key' so that the paint film can adhere strongly to the surface of the work, but also to render the surface under the paint film chemically neutral. This ensures that no corrosive or other reactions take place which might attack the component under the paint film. The products of such reactions lift the paint film causing it to bubble and flake, thus breaking down its protection.

1.33 Application of paints

The simplest way to apply paint is by brushing it over the surface to be protected. However, it is also the most labour-intensive and costly way of painting any surface. Further, there is little control of the quality of the film so applied. For manufacturing purposes where large quantities of the same product are to be painted and where low cost coupled with uniform quality is of paramount importance, more sophisticated methods of application such as dipping, conventional spraying (compressed air), airless spraying, and electrostatic spraying (conventional and airless), have to be adopted.

1.34 Plastic degradation

Whilst plastic materials are highly resistant to atmospheric attack, even in urban and marine environments, they degrade to the point of destruction very rapidly in the presence of the ultra-violet rays found in sunlight. For this reason colouring agents have to be included in the

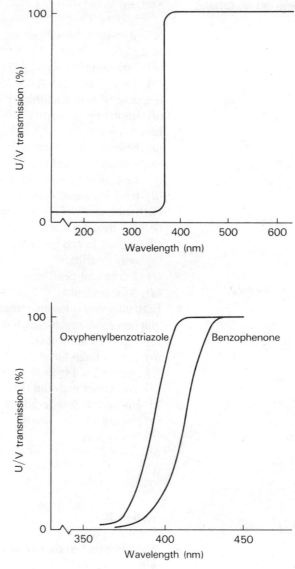

(a) **Transmission curve for an *ideal* ultra-violet light absorbtion additive**

(b) **Transmission curves for *actual* ultra-violet light absorbtion additives**

Fig. 1.28 Ultra-violet light absorbtion additives

plastic to filter out these rays. Carbon black is an ultra-violet ray absorber but gives a black product. Alternative and highly effective additives which absorb the ultra-violet rays whilst allowing the use of more attractive colourants are *oxyphenylbenzotriazole* and *benzophenone*, the absorption properties of which are shown in Fig. 1.28. The presence of inorganic solvent fumes in the atmosphere can reduce the tensile strength of most plastic materials. Because of the difficulty in destroying unwanted plastic materials except by burning, some are now being produced which are 'bio-degradable' and are destroyed by the bacteria found in the ground and on rubbish dumps.

Problems (Engineering Materials)   **Section A**

1 Non-metallic components are used in electrical equipment because of their:
   (*a*) light weight;
   (*b*) good conductivity;
   (*c*) ease of manufacture;
   (*d*) insulating properties.

2 Low carbon steel that has been shaped by hot-working tends to:
   (*a*) become work hardened;
   (*b*) remain soft and ductile;
   (*c*) have an improved surface finish;
   (*d*) have increased elasticity.

3 High carbon (hyper-eutectoid) steels can be used for cutting tools because they contain:
   (*a*) cementite and pearlite;
   (*b*) only pearlite;
   (*c*) ferrite and pearlite;
   (*d*) only cementite.

4 High duty cast irons are stronger than grey cast iron because the free graphite (uncombined carbon) is:
   (*a*) removed altogether;
   (*b*) left in flake form;
   (*c*) spherodised by adding traces of magnesium or cerium;
   (*d*) combined with the iron to form cementite.

5 In 'gun metal' bronze alloys the de-oxidising agent is:
   (*a*) copper;
   (*b*) phosphorus;
   (*c*) zinc;
   (*d*) tin.

Section B

6 With reference to the properties of the materials concerned, explain why:
   (*a*) the overhead electrical transmission lines of the 'grid system' have a high tensile steel wire core and an outer sheath of pure aluminium wires;
   (*b*) soldering irons have copper 'bits';
   (*c*) beryllium copper tools are frequently used on oil drilling rigs and down coal mines.

7 Explain why laminated plastic (Tufnol) is highly suitable for sea-water pump bearings. Your answer should consider such factors as:
   (*a*) electrolytic corrosion;
   (*b*) mechanical strength;
   (*c*) coefficient of friction;
   (*d*) lubrication.

8 List the essential properties of a metallic-bearing alloy and explain how 'Babbitt' metal fulfils these requirements.

9 (a) State the difference between thermo-plastic and thermosetting plastic materials;

(b) Select a suitable plastic material for each of the following applications, giving reasons for your choice: (i) the insulation of flexible cables; (ii) the rigid casing for an ammeter; (iii) a safety helmet; (iv) a heavy duty gear wheel; (v) a light duty bearing bush for an office machine; (vi) a transparent moulded cover for an instrument.

10 State a typical composition for each of the following non-ferrous alloys and give a typical application, with reasons for your choice, in each case:

(a) cartridge brass;

(b) free-cutting brass;

(c) admiralty brass;

(d) phosphor bronze;

(e) gun-metal bronze;

(f) silver copper;

(g) duralumin;

(h) tinman's solder.

Problems (Heat Treatment)

Section A

1 The correct hardening temperature for a 0.6% carbon chisel steel is:

(a) slightly above its lower critical temperature;

(b) slightly above its upper critical temperature;

(c) slightly below its lower critical temperature;

(d) as high as possible without melting the steel.

2 The purpose of tempering a hardened steel component is to:

(a) increase its toughness;

(b) reduce its hardness;

(c) increase its hardness;

(d) increase its ductility.

3 The solution treatment of aluminium alloys consists of:

(a) raising the temperature of the alloy so that the elements can form a solid solution;

(b) heating the alloy in a solution of chemicals;

(c) melting the alloy and adding other solids to form a solution;

(d) a corrosion proofing process.

4 The precipitation treatment of aluminium alloys is used to:

(a) soften the alloys;

(b) harden the alloys;

(c) increase the density of castings;

(d) remove impurities from the alloys.

5 To achieve the maximum hardness when quench-hardening high carbon (hyper eutectoid) steels it is essential to:

(*a*) raise the temperature of the steel as high as possible;
(*b*) soak the steel at a high temperature for as long as possible;
(*c*) cool the steel slowly in oil;
(*d*) cool the steel quickly in water.

Section B

6 A batch of work-hardened mild steel pressings are to be sub-critically annealed. Describe:
(*a*) what is meant by the term *sub-critical annealing*;
(*b*) how the process is carried out;
(*c*) the advantages of this process over full annealing in this instance.
7 Describe the essential differences between *normalising* and full *annealing* and give an example of where each process would be used.
8 Describe the difference between hot- and cold-working when flow-forming metal components and compare the advantages and limitations of each process.
9 Explain, in detail, how a chisel made from a silver steel (high carbon) steel rod should be hardened and tempered for cutting cast iron.
10 (*a*) Explain how duralumin rivets should be softened.
(*b*) Explain how the process of 'age-hardening' may be delayed after the rivets have been softened.
(*c*) Explain, briefly, the cause of age-hardening in duralumin alloys.

Problems (Corrosion)

Section A

1 The 'rusting' of steel is caused by the action of:
(*a*) air (oxygen) alone;
(*b*) moisture alone;
(*c*) air and moisture together;
(*d*) industrial pollutants.
2 The most effective method of preventing the atmospheric corrosion of steels is by:
(*a*) galvanising;
(*b*) anodising;
(*c*) greasing;
(*d*) painting.
3 Copper is protected from atmospheric corrosion by the formation of a surface film called:
(*a*) verdigris;
(*b*) copper oxide;
(*c*) copper sulphate;
(*d*) patina.
4 Aluminium alloys have their corrosion resistance improved by:
(*a*) galvanising;
(*b*) chromising;

(*c*) calorising;

(*d*) anodising.

5 Phosphate anti-corrosion coatings are used to prepare metals for:

(*a*) painting;

(*b*) plastic coating;

(*c*) metal spraying;

(*d*) electro-plating.

Section B

6 (*a*) Zinc is used to galvanise steel components. Explain what is meant by the zinc coating being 'sacrificial'.

(*b*) Calculate the mass of nickel deposited on the cathode of an electro-plating cell if a current of 5 amperes flows for 3 hours and the cell is 85 per cent efficient. (The electro-chemical equivalent of nickel is 0.304.)

7 Describe the precautions that must be taken when:

(*a*) paint spraying;

(*b*) stove enamelling.

8 Paints are classified by the way in which they dry or harden. Describe the four different systems available and state a typical application of each system.

9 (*a*) Explain with the aid of a diagram how the gases given off by factory chimneys form corrosive pollutants in the atmosphere and list the more destructive of them.

(*b*) Explain why it is always necessary to prepare a metal surface carefully before applying anti-corrosion treatment.

(*c*) Describe how structural steelwork can be prepared prior to painting.

10 Explain how anodising protects aluminium and aluminium alloys and how it differs fundamentally from electro-plating.

2 Engineering drawing

2.1 The need for orthographic drawing

A well drawn picture or series of pictures can describe an object more quickly and accurately than a written description. Moreover there is less chance of misinterpretation. Figure 2.1 shows a photograph of a typical engineering component. To draw this solid object on a flat sheet of paper special techniques have to be used. These techniques are called *projections*. There are many different projections available to the draftsperson and some examples are shown in Fig. 2.2. The pictorial drawings shown in Fig. 2.2 are easy to understand and the solid object they represent is easy to visualise. However, they are all distorted to some extent to give the illusion of 'depth'.

The engineer is not concerned with the artistic merit of a drawing, only with the information it communicates and whether this information is accurate and without the possibility of misunderstanding. Therefore, although the pictorial representations shown are excellent for operators' handbooks, maintenance manuals, and similar applications, where the true size is unimportant, engineering drawings for manufacturing and assembly purposes are produced by a technique known as *orthographic projection*. Figure 2.3 shows a typical *orthographic* drawing of the component shown in Fig. 2.1. It can be seen that no attempt has been made at pictorial realism. In an orthographic drawing, three separate drawings or 'views' of the component are drawn on the same sheet of paper but from different viewpoints. Simple components can have fewer than three views and complex components may need more than the three basic views shown. Orthographic drawing allows each view to be drawn true to scale without distortion and, with practice, the drawings can be quickly, easily and accurately interpreted. The principles of orthographic projection will now be considered.

2.2 Orthographic projection (first angle)

Orthographic projection uses two main projection planes, called *principal planes*. They intersect each other at right-angles, as shown in Fig. 2.4, to form four quadrants. The component to be drawn is placed

Fig. 2.1 Vice body

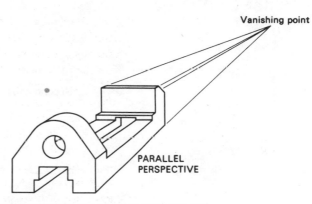

Vanishing point

PARALLEL
PERSPECTIVE

*THIS TECHNIQUE IS RARELY USED BY THE ENGINEERING
DRAUGHTSPERSON*

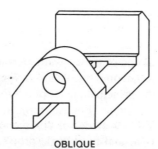

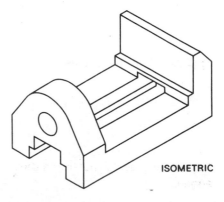

ISOMETRIC

Fig. 2.2 Techniques of drawing —
pictorial

OBLIQUE

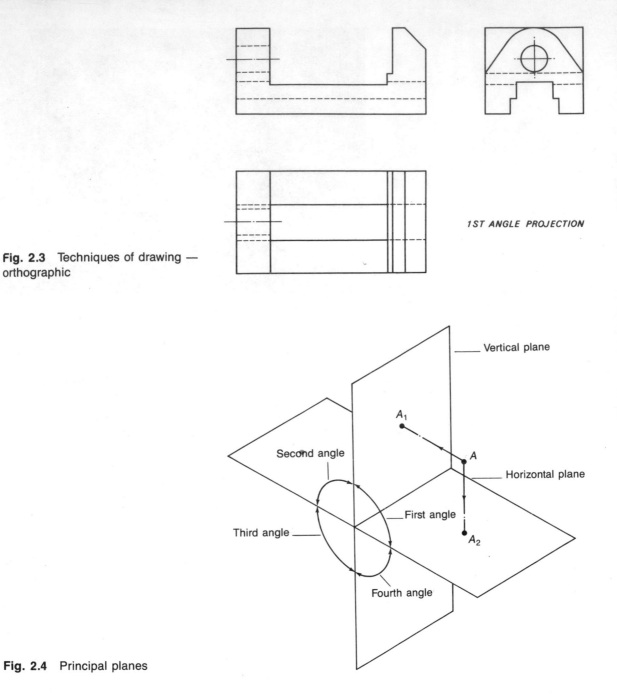

1ST ANGLE PROJECTION

Fig. 2.3 Techniques of drawing — orthographic

Fig. 2.4 Principal planes

in one of these quadrants and then orthographic views of it are projected onto the planes. In Fig. 2.4 the point *A* has been projected onto the principal planes of the *first angle*. In practice, engineers only use *first angle* (English) projection or third angle (American) projection. The remainder of this section will only consider first angle projection, and third angle projection will be considered in section 2.3.

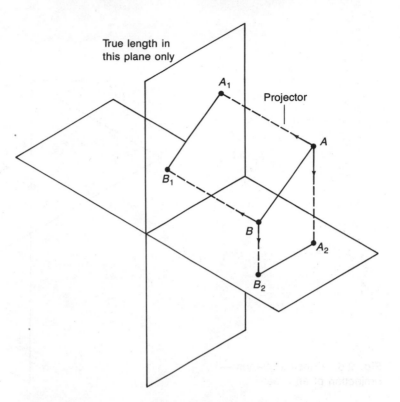

Fig. 2.5 Principal planes —
projection of a line

Figure 2.5 shows a *line* projected onto the principal planes. This
is done by projecting its end onto the plane using 'projectors' as shown.
The projectors are parallel to each other and perpendicular (at right-
angles) to the plane. It can be seen that the line can only appear as
having a true length if it lies parallel to the plane.

Figure 2.6 shows an *area* projected onto the principal planes. In this
instance an area is defined by points connected by lines. Thus, by projecting the points at the ends of the lines onto the principal planes,
the area can be projected onto the planes. Again, the area will only
appear true-size on a plane which is parallel to it. Figure 2.7 shows
how an *auxiliary plane* can be used to get true length lines and,
therefore, a true area.

The projection of a *solid object* onto the principal planes is shown
in Fig. 2.8. In order to accommodate all the views of the object it has
been necessary to provide an additional *auxiliary plane*. Views of the
object have been drawn on the planes using parallel projectors as
previously. The view on the vertical plane is called the *elevation*; that
on the auxiliary vertical plane the *end elevation* or the *end view*; that
on the horizontal plane is called the *plan*. To obtain the views as they
would appear on a sheet of paper, the planes have to be opened out,
or rabatted, about the intersection of the planes as shown in Fig. 2.9.
The line forming the intersection of the principal planes is called the
ground line.

In practice, the drawing is constructed as shown in Fig. 2.10. The

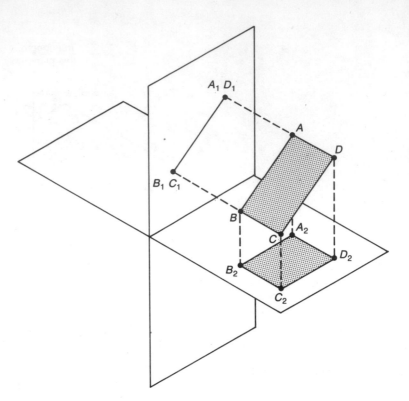

Fig. 2.6 Principal planes —
projection of an area

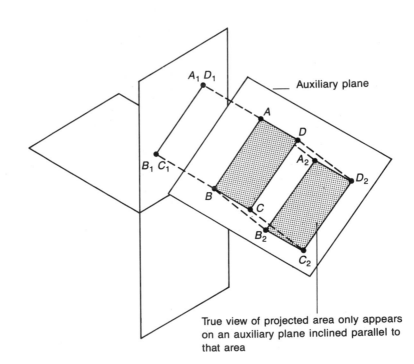

Fig. 2.7 Auxiliary plane

Auxiliary plane

True view of projected area only appears
on an auxiliary plane inclined parallel to
that area

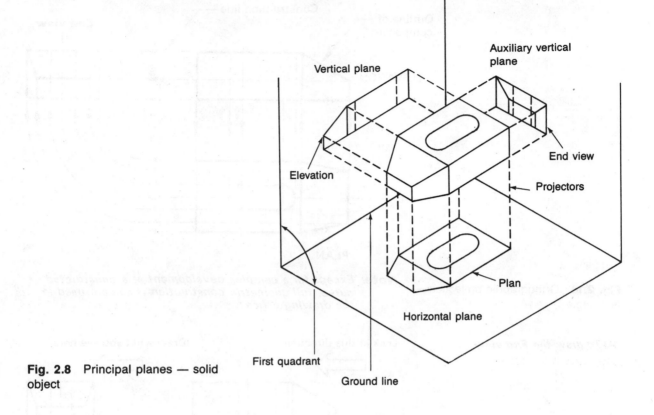

Fig. 2.8 Principal planes — solid object

Labels: Vertical plane, Auxiliary vertical plane, End view, Projectors, Elevation, Plan, Horizontal plane, First quadrant, Ground line

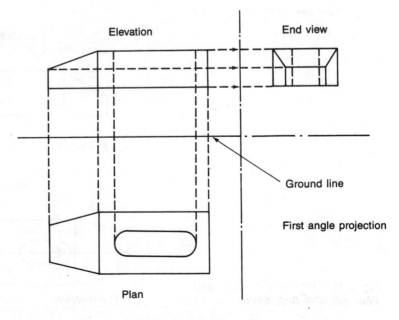

Labels: Elevation, End view, Ground line, First angle projection, Plan

Fig. 2.9 Rebatment of principal planes — first angle

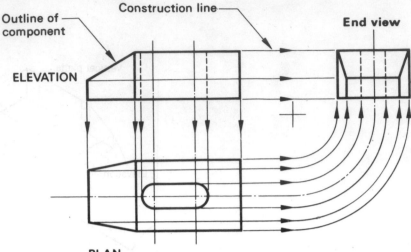

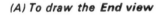

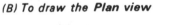

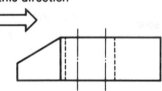

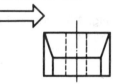

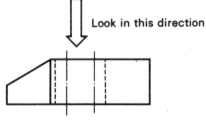

Fig. 2.10 Orthographic projection — first angle

Note: Except for a complex development or a constructed curve, full geometric construction is seldom used in the drawing office.

(A) To draw the End view

Look in this direction   Draw what you see here

ELEVATION (side view)   (end view)

(B) To draw the Plan view

Look in this direction

ELEVATION (side view)   (end view)

Draw what you see here

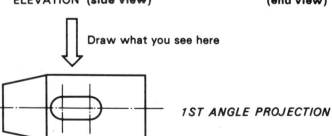

1ST ANGLE PROJECTION

PLAN VIEW

Fig. 2.11 Summary of views and view points — first angle

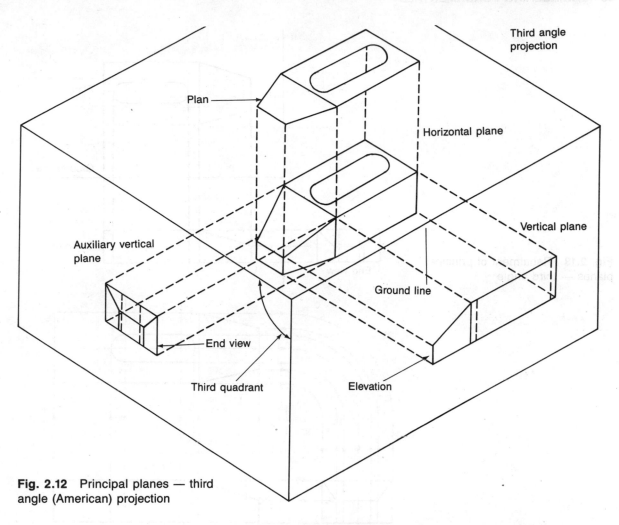

Third angle
projection

Plan

Horizontal plane

Auxiliary vertical
plane

Vertical plane

Ground line

End view

Third quadrant

Elevation

Fig. 2.12 Principal planes — third
angle (American) projection

lines used to project the points of the component are called *construction lines* and they should be feint, and thin. The points where they intersect indicate the corners of the component. When all the construction lines are in place, the corner points are joined up with heavier lines to indicate the outline of the component. The relationship between view points and the position of the drawn views for first angle projection is summarised in Fig. 2.11.

2.3 Orthographic projection (third angle)

This time the component originally shown in Fig. 2.8 has been repositioned in the *third quadrant* as shown in Fig. 2.12. Since the planes now lie between the view point and the component, they have to be considered as transparent. When the planes are rabatted as shown in Fig. 2.13, the plan is now above the elevation and the end view is at the same end of the component from which it is viewed.

In practice, the drawing is constructed as shown in Fig. 2.14. Again, the construction lines should be feint and thin. The points where they

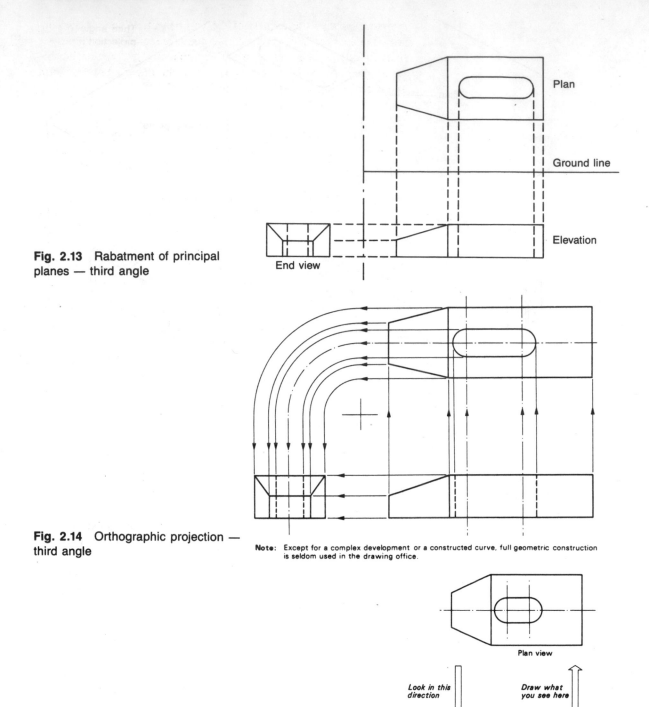

Fig. 2.13 Rabatment of principal planes — third angle

Fig. 2.14 Orthographic projection — third angle

Note: Except for a complex development or a constructed curve, full geometric construction is seldom used in the drawing office.

Fig. 2.15 Summary of views and view points — third angle

intersect indicate the corners of the component. When all the construction lines are in place, the corner points are joined up with heavier lines to indicate the outline of the component. The relationship between the view points and the position of the drawn views for third angle projection is summarised in Fig. 2.15.

2.4 Auxiliary view

The use of an auxiliary plane to give a true view of an inclined area was introduced in Fig. 2.7. This technique is important in the production of working drawings so that the true shape of an inclined surface can be seen and so that the position of features on the surface can be dimensioned to fix their positions. Figure 2.16 shows a bracket with an inclined face. When it is drawn in first-angle projection it can be seen that the inclined surface and its features are heavily distorted, but they appear correct in size and shape in the *auxiliary view* (AV) which is projected *at right-angles* (perpendicular) to the inclined face.

2.5 Oblique projection

At the start of this chapter, Fig. 2.2 showed how a solid object could be represented on a flat sheet of paper using various 'pictorial' pro-

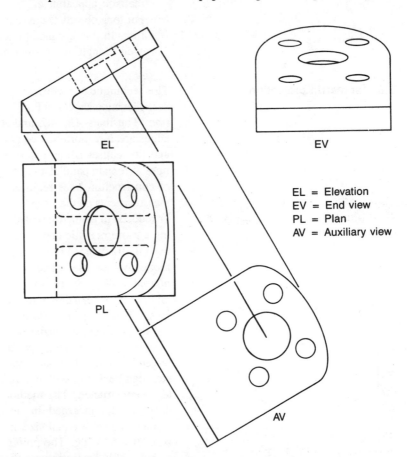

EL = Elevation
EV = End view
PL = Plan
AV = Auxiliary view

Fig. 2.16 Auxiliary view

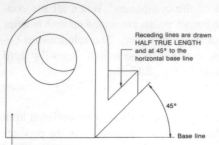

Receding lines are drawn
HALF TRUE LENGTH
and at 45° to the
horizontal base line

45°

Base line

The elevation is drawn square on
as in orthographic drawing. All
circles and arcs can be drawn with
compasses. All line are
TRUE LENGTH

Fig. 2.17 Oblique projection

jection techniques. The simplest of these is the *oblique* projection technique as shown in Fig. 2.17. The component is positioned so that one face can be drawn as its true shape and size. That is, two of the axes are at right angles to each other. The third axis, known as the *receding* axis may be at any angle but is usually set at 30° or at 45° so that standard set squares can be used to draw the parallel projection lines. When the angle chosen is 45°, the drawing is said to be in 'cavalier projection'. Since no allowance is made for perspective in the receding lines, the component looks distorted. To overcome this, the dimensions on the receding axis can be halved. This is called 'cabinet projection'.

For ease of drawing and realism, there are some simple rules for setting out the drawing.

(*a*) Any curved or irregular face should be drawn true shape. For example, a circle on a receding face would have to be constructed as an ellipse, whereas on the front face it can be drawn true shape with compasses.

(*b*) Wherever possible the longest side should be shown on the front, true view. This prevents violation of perspective and gives a more realistic appearance.

(*c*) For long objects of circular cross-section, the above two rules conflict. In this instance the circular shape should be the true view even though this results in the long axis receding.

2.6 Isometric projection

This technique of pictorial representation was also introduced in Fig. 2.2. The cube shown in Fig. 2.18 has been drawn in *isometric projection*. The lines *AB, AC*, and *AD* are the isometric axes and all the edges are foreshortened equally. No account is taken of perspective and this causes some distortion. However it does allow for parallel lines to remain parallel and this eases drawing and reduces the number of construction lines required. When making an isometric drawing measurements on the isometric lines can be true (natural scale) length, or reduced to isometric scale. Although the latter scale gives the drawing a more realistic appearance, it is not so convenient and is rarely used in practice. In this chapter only true length (natural scale) techniques will be used. The cube shown in Fig. 2.18 has been drawn to natural scale. A more complex component is shown in Fig. 2.19 and all lines have been drawn true length. When producing pictorial drawings, it is not usual to show hidden detail unless this is absolutely essential to establish the shape of the component.

Although this technique produces a better representation than oblique projection, it has the disadvantage that no curved profiles (circles, radii, etc.) can be drawn with compasses and all such curved profiles have to be constructed. The method of doing this is to erect a grid over the feature concerned in orthographic projection, then draw an isometric grid of equal size at the point where it is to appear on the isometric drawing. The points where the curve cuts the grid in the orthographic projection are then transferred to the isometric grid by

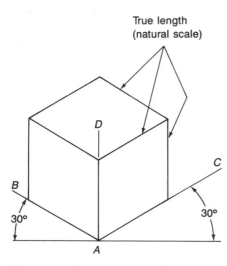

True length
(natural scale)

D

C

B

30°   30°

A

Fig. 2.18 Isometric projection

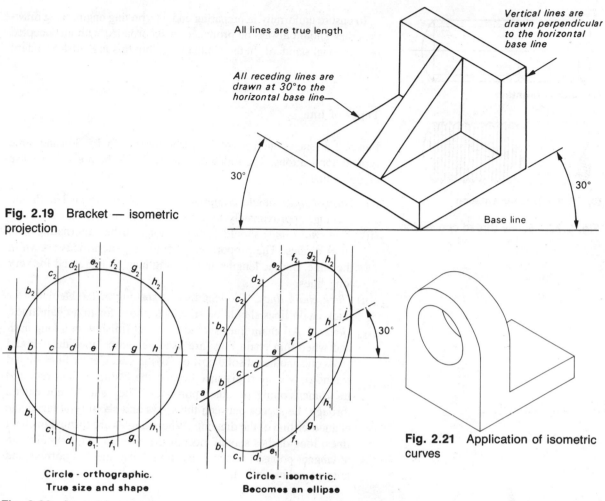

All lines are true length

All receding lines are drawn at 30° to the horizontal base line

Vertical lines are drawn perpendicular to the horizontal base line

30°

30°

Base line

Fig. 2.19 Bracket — isometric projection

Circle - orthographic.
True size and shape

Fig. 2.20 Isometric curves — construction

Circle - isometric.
Becomes an ellipse

Fig. 2.21 Application of isometric curves

stepping off their true lengths from the axis as shown in Fig. 2.20. Figure 2.21 shows Fig. 2.17 redrawn in isometric projection with the curves now represented by ellipses or part ellipses.

2.7 Conventions

The engineering drawing is only a means of recording the intentions of the designer and communicating those intensions to the manufacturer. It is not a work of art and, apart from the time expended in its preparation, is of no intrinsic value. If a better and cheaper method of communication could be discovered, then engineering drawings would no longer be used. Therefore the engineering drawing must be prepared as quickly and easily as possible to keep costs down. To this end, *standard conventions* are used to help speed the drafting (drawing) process by representing the more frequently used and awkward details in a form of universally understood drawing 'shorthand'. For example, the representation of a screw thread shown in Fig. 2.22(*a*) is much easier to draw than its pictorial representation shown in Fig. 2.22(*b*).

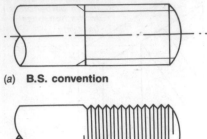

(a) **B.S. convention**

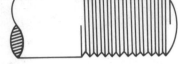

(b) **Pictorial representation**

Fig. 2.22 Screw thread convention

To ensure uniformity in preparing and interpreting engineering drawings they should always be produced in accordance with an accepted international standard. In the United Kingdom this is British Standard 308.

Types of line

Figure 2.23 shows the types of line recommended by BS 308 and some typical applications. The following points should be noted in the use of these lines.

(i) *Dashed lines* should comprise dashes of consistent length and spacing, approximately to the proportions in the table.

(ii) *Thin chain lines* should comprise long dashes alternating with short dashes. The proportions should be generally as shown in the table, but the lengths and spacing may be increased for very long lines.

(iii) *Thick chain lines.* The lengths and spacing of the elements for thick chain lines should be similar to those for thin chain lines.

(iv) *General.* All chain lines should start and finish with a long dash and when thin chain lines are used as centre lines they should cross one another at solid portions of the line. Centre lines should extend only a short distance beyond the feature unless required for dimensioning or other purposes. They should not extend through the spaces between the views and should not terminate at another line of the drawing. Where angles are formed in chain lines, long dashes should meet at the corners. Arcs should join at tangent points. Dashed lines should also meet at corners and tangent points with dashes.

Conventions for common features

Figure 2.24 shows some typical conventions used in engineering drawings. It is not possible, within the scope of this book, to provide the full set of conventions or to provide detailed explanations of their applications. For this it is necessary to consult texts specialising in engineering drawing together with the full British Standard for Engineering Drawing Practice (BS 308).

Abbreviations for written statements

Table 2.1 lists the standard abbreviations for written statements as used on engineering drawings.

2.8 Redundant views

A ball looks the same from every direction, and to represent it by three circles all the same size and carefully arranged as elevation, end view

Line	Description	Application
A	Continuous thick	A1 Visible outlines A2 Visible edges
B	Continuous thin	B1 Imaginary lines of intersection B2 Dimension lines B3 Projection lines B4 Leader lines B5 Hatching B6 Outlines of revolved sections B7 Short centre lines
C	Continuous thin irregular	*C1 Limits of partial or interrupted views and sections, if the limit is not an axis
D	Continuous thin straight with zigzags	†D1 Limits of partial or interrupted views and sections, if the limit is not an axis
E	Dashed thick	E1 Hidden outlines E2 Hidden edges
F	Dashed thin‡	F1 Hidden outlines F2 Hidden edges
G	Chain thin	G1 Centre lines G2 Lines of symmetry G3 Trajectories and loci G4 Pitch lines and pitch circles
H	Chain thin, thick at ends and changes of direction	H1 Cutting planes
J	Chain thick	J1 Indication of lines or surfaces to which a special requirement applies (drawn adjacent to surface)
K	Chain thin double dashed	K1 Outlines and edges of adjacent parts K2 Outlines and edges of alternative and extreme positions of movable parts K3 Centroidal lines K4 Initial outlines prior to forming §K5 Parts situated in front of a cutting plane K6 Bend lines on developed blanks or patterns

NOTE. The lengths of the long dashes shown for lines G, H, J and K are not necessarily typical due to the confines of the space available.

*See also 8.8 and 9.7.

†This type of line is suited for production of drawings by machines.

‡The thin F type line is more common in the UK, but on any one drawing or set of drawings only one type of dashed line should be used.

§Included in ISO 128-1982 and used mainly in the building industry.

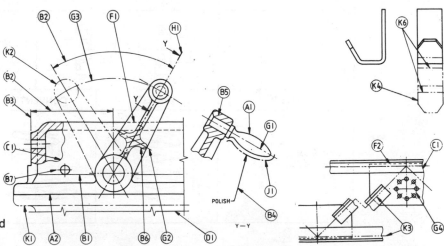

Fig. 2.23 Types of line and their applications

and plan view would be a waste of time. One circle and a note that the component is spherical is all that is required. The views which can be discarded without loss of information are called *redundant views*. Figures 2.25 and 2.26 show how drawing time can be saved and the drawing simplified by eliminating the redundant views when drawing symmetrical components.

Fig. 2.24 Conventions for some common features

TITLE	SUBJECT	CONVENTION	
External screw threads (detail)			
Internal screw threads (detail)			
Diamond knurling			
Square on shaft			
Holes on circular pitch			
Bearings			
TITLE	SUBJECT	CONVENTION	DIAGRAMMATIC REPRESENTATION
Cylindrical compression spring		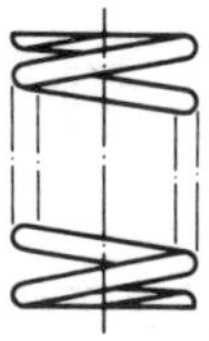	

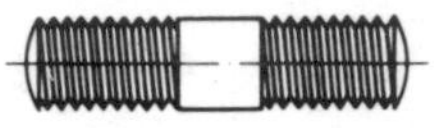

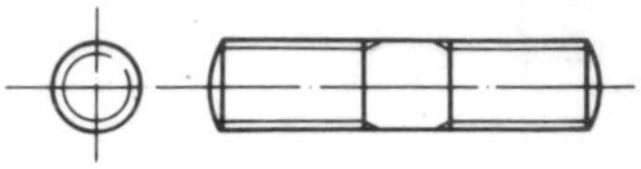

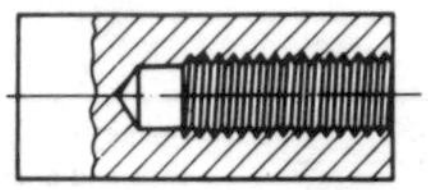

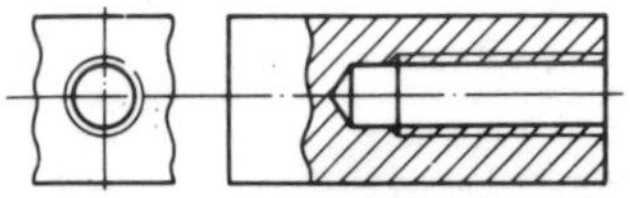

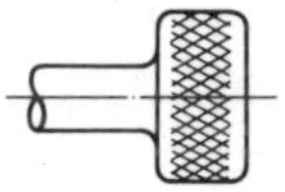

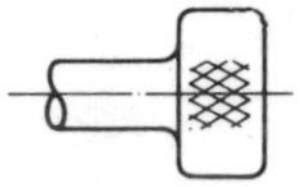

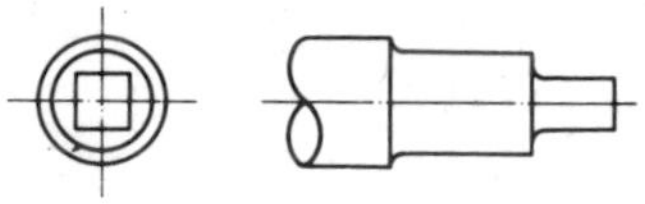

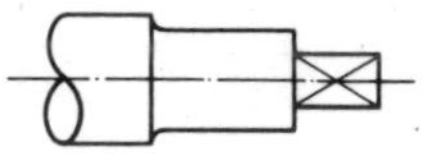

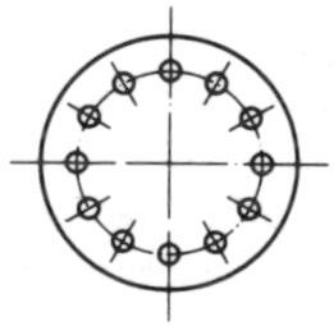

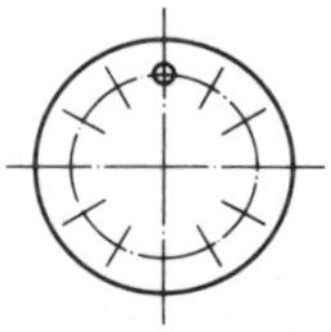

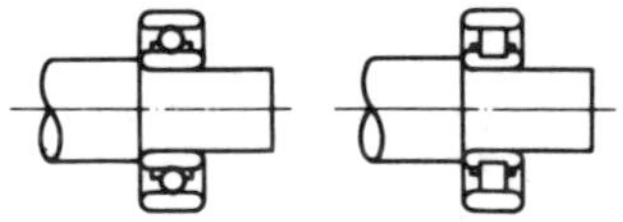

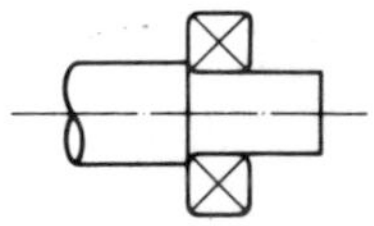

Table 2.1 Abbreviations for written statements

Term	Abbreviation
Across flats	A/F
British Standard	BS
Centres	CRS
Centre line	CL or ℄
Chamfered	CHAM
Cheese head	CH HD
Countersunk	CSK
Countersunk head	CSK HD
Counterbore	C'BORE
Diameter (in a note)	DIA
Diameter (preceding a dimension)	Ø
Drawing	DRG
Figure	FIG.
Hexagon	HEX
Hexagon head	HEX HD
Material	MATL
Number	NO.
Pitch circle diameter	PCD
Radius (in a note)	RAD
Radius (preceding a dimension)	R
Screwed	SCR
Specification	SPEC
Spherical diameter or radius	SPHERE Ø or R
Spotface	S'FACE
Standard	STD
Undercut	U'CUT

2.9 Dimensioning

So far only the shape of the component has been considered. However in order that the component can be manufactured, the drawing must also show the size of the component and the position and size of any features on the component. To avoid confusion and the chance of misinterpretation, the dimensions which prescribe the size of the component and its features must be added to the drawing in the manner laid down by BS 308. Figure 2.27(*a*) shows how projection and dimension lines are used to relate the dimension to the drawing, whilst Fig. 2.27(*b*) shows the correct and incorrect methods of dimensioning a drawing.

Correct dimensioning

(*a*) Dimension lines should be thin full lines not more than half the thickness of the component outline (see Fig. 2.23).

(*b*) Wherever possible dimension lines should be placed outside the outline of the drawing.

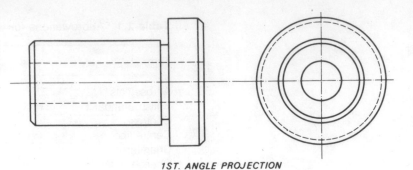

1ST. ANGLE PROJECTION

Fig. 2.25   Cylindrical components

*Note: When making an Orthographic drawing of a symmetrical component, such as this circular bush, it is not essential to provide **three** views.*

(*c*) The dimension line arrowhead must touch the projection line.

(*d*) Dimension lines should be well spaced so that the numerical value of the dimension can be clearly read and also so that they do not obscure the outline of the drawing.

Incorrect dimensioning

(*a*) Centre lines and extension lines must not be used as dimension lines.

(*b*) Wherever possible dimension line arrowheads must not touch the outline directly, but should touch the projection lines which extend from the outline.

(*c*) If the use of a dimension line within the outline is unavoidable, then try and use a leader line to take the dimension itself outside the outline.

Figure 2.28(*a*) shows how circles and shaft ends (diameters) should be dimensioned.

It is preferable to use those techniques which take the dimension outside the circle unless the circle is so large that the dimension will neither be cramped and difficult to read nor will it obscure any vital feature. Note the use of the symbol to denote a diameter.

Figure 2.28(*b*) shows how radii are dimensioned. Note that the radii of arcs of circles need not have their centres located if the start and finish points are known.

Figure 2.28(*c*) shows how notes may be used to avoid the need for the full dimensioning of certain features of a drawing.

Figure 2.29 shows how leader lines indicate where notes or dimensions are intended to apply. Such lines are terminated with arrowheads or dots. *Arrowheads* are used where the leader line touches the outline of the component or feature. Dots are used where the leader line finishes within the outline of the feature to which it refers.

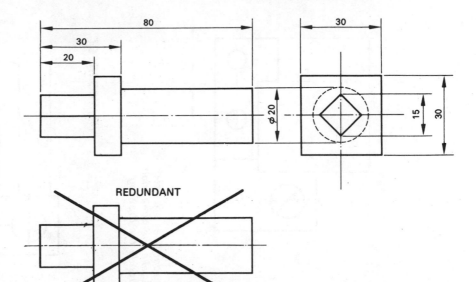

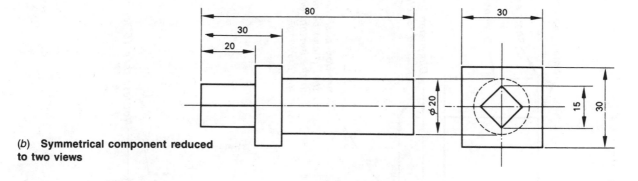

(a) **Working drawing of a symmetrical component in 1st angle. Plan view redundant**

(b) **Symmetrical component reduced to two views**

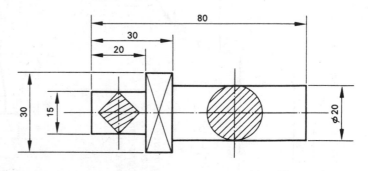

(c) **Working drawing of (b) reduced to single view by using revolved sections and B.S. convention for the square flange.**

Fig. 2.26 Redundant views

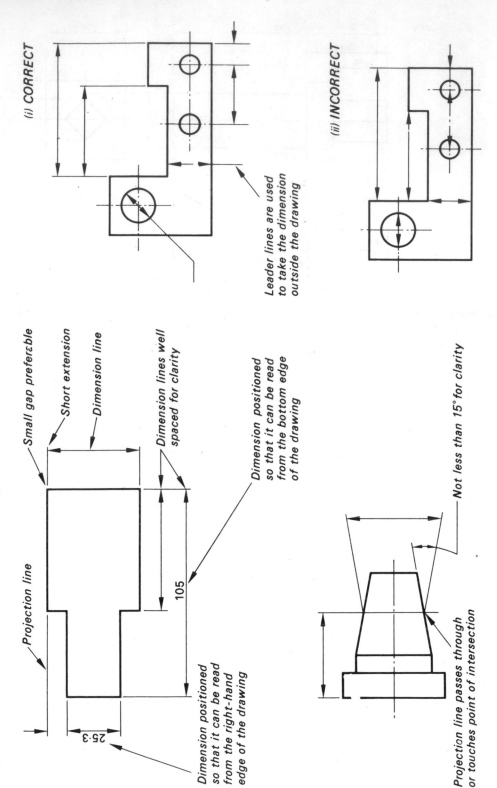

(i) CORRECT

Leader lines are used to take the dimension outside the drawing

(iii) INCORRECT

(b) **Correct and incorrect dimensioning**

Small gap preferable

Short extension

Dimension line

Dimension lines well spaced for clarity

Dimension positioned so that it can be read from the bottom edge of the drawing

105

Projection line

Dimension positioned so that it can be read from the right-hand edge of the drawing

25·3

Not less than 15° for clarity

Projection line passes through or touches point of intersection

(a) **Projection and dimension lines**

Fig. 2.27 Dimensioning

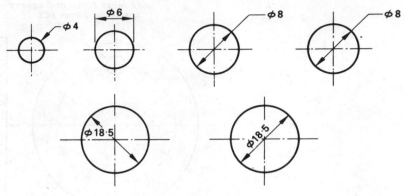

(a) **Dimensioning holes**

Note : *Leader to be in line with hole centre*

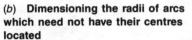

(b) **Dimensioning the radii of arcs which need not have their centres located**

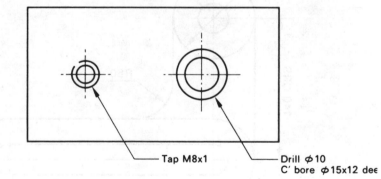

(c) **Use of notes to save full dimensioning**

Fig. 2.28 Dimensioning — diameters and radii

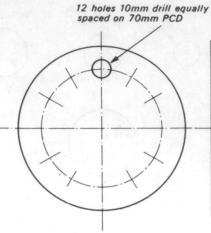

12 holes 10mm drill equally
spaced on 70mm PCD

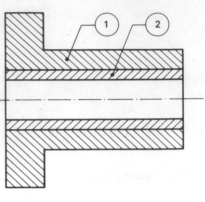

Fig. 2.29 Termination of leader
lines

(a) **Leader lines terminating
in arrowheads**

(b) **Leader lines terminating
in dots**

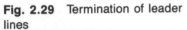

Dimensions in millimetres

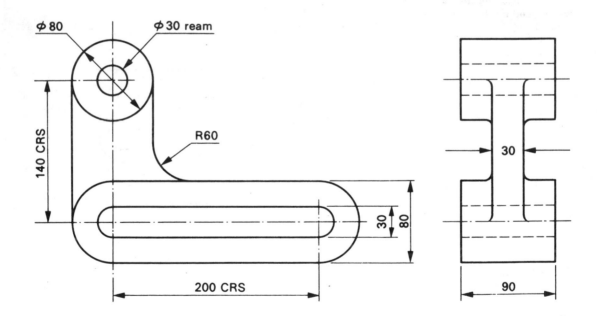

Dimensions should be placed in the view which shows the feature in clear outline

Fig. 2.30 Correct positioning of
dimensions

2.10 Duplication and selection of dimensions

In the same way as the number of views in a drawing are kept to a minimum, dimensions are selected to avoid confusion by unnecessary detail. Duplication of dimensions must also be avoided. The dimensions should appear adjacent to the view in which the feature being dimensioned appears as a true shape and size. For example, a hole should be dimensioned in the view in which it appears as a circle and not where it appears as hidden detail. Figure 2.30 shows how dimensions should be correctly positioned.

2.11 Auxiliary dimensions

To avoid mistakes, it has already been stated that duplicated or unnecessary dimensions should not appear on a drawing. The only exception to this rule is when *auxiliary dimensions* are used to avoid the calculation of, say, overall dimensions on the shop floor. Such dimensions are placed in brackets as shown in Fig. 2.31. The dimensions marked *F* are the 'functional' dimensions from which the component is manufactured. Auxiliary dimensions are also referred to as *non-functional* dimensions.

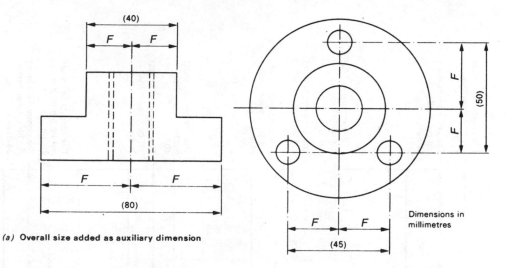

(a) Overall size added as auxiliary dimension

(b) Hole centres added as auxiliary dimension

Dimensions in millimetres

Fig. 2.31 Auxiliary dimensions

2.12 Toleranced dimensions

It is true to say that if a component was ever made exactly to size no one would ever know because it could not be measured exactly. Having calculated the ideal size for a dimension, the designer must then decide how much variation from that size he will tolerate. This variation between the smallest and the largest acceptable size is called the *tolerance*. As well as specifying the magnitude of the tolerance the designer must also indicate where the tolerance lies relative to the nominal size. Figure 2.32(a) shows various methods of tolerancing a dimension and how the tolerance is interpreted. Figure 2.32(b) shows the types of 'fit' which can be achieved between mating components, whilst Fig. 2.32(c) explains the terms used.

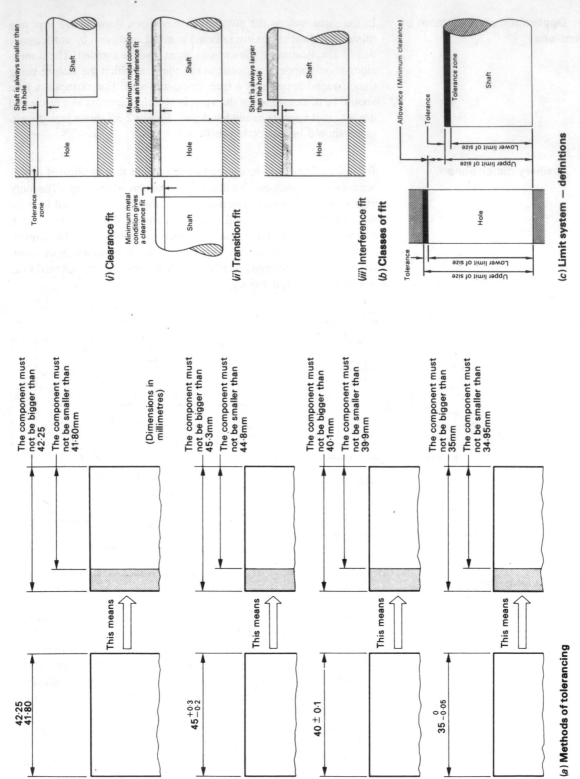

(i) Clearance fit

(ii) Transition fit

(iii) Interference fit

(b) Classes of fit

(c) Limit system — definitions

(a) Methods of tolerancing

Fig. 2.32 Toleranced dimensions

When toleranced dimensions are used, cumulative errors can occur wherever a feature is controlled by more than one tolerance as shown in Fig. 2.33(*a*). It can be seen that chain dimensioning gives a build up of tolerances which is greater than the designer intended. In this example the maximum tolerance for the right-hand hole centre is ±0.3 mm from the left hand datum edge. This cumulative effect can easily be eliminated by dimensioning each feature from a common datum as shown in Fig. 2.33(*b*).

Cumulative errors can also occur when a number of components are assembled together as shown in Fig. 2.34. To avoid this, assembly dimensions are given. The individual components are adjusted within their individual tolerances, or selective assembly is used so the cumulative error is within the tolerance of the assembly dimensions. In the example shown, the dimension *A* of the assembled components

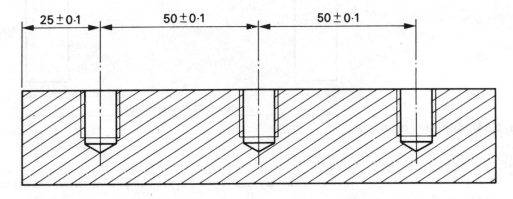

(a) **String dimensioning. Cumulative tolerance equals sum of individual tolerances**

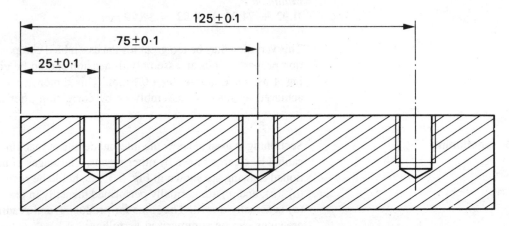

(b) **Dimensioning from one common datum to eliminate cumulative effect**

Fig. 2.33 Cumulative error

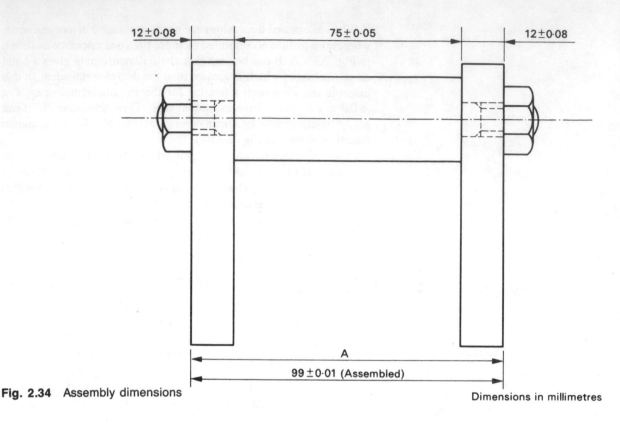

Fig. 2.34 Assembly dimensions

Dimensions in millimetres

can vary as follows for the component tolerances given:

Maximum
12.08 + 75.05 + 12.08 = *99.21 mm*

Minimum
11.92 + 74.95 + 11.92 = *98.79 mm*

This variation may be too great for the assembled components to function properly. Thus an assembly dimension should be given specifying *A* as, for example, 99 ± 0.1 mm. This dimension could then be achieved by selective assembly or by correction after assembly.

2.13 Sectioning

Sectioning is used to show the internal details of machine parts which cannot be clearly shown by other means. The stages of making a sectioned drawing are shown in Fig. 2.35. It should be realised that steps (*a*), (*b*), and (*c*) are normally performed mentally in practice and only (*d*) is actually drawn. The rules for producing and reading sectioned drawings can be summarised as follows.

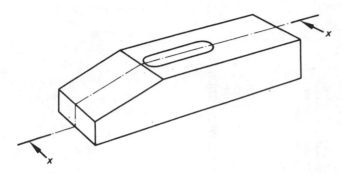

(a) The clamp is to be Sectioned along the line xx

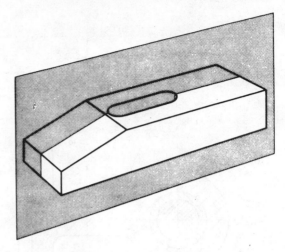

(b) The cutting plane is positioned on the line xx as shown

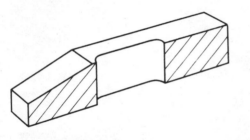

(c) That part of the clamp that lies in front of the cutting plane is removed leaving the sectioned component

Fig. 2.35 Section drawing

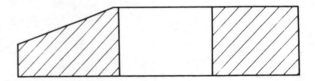

(d) Sectioned, orthographic elevation of the clamp shown in (a).

Note the section shading lines lie at 45° to the horizontal and are half the thickness of the outline.

(*a*) Drawings are only sectioned when it is impossible to show the internal detail of a component in any other way.

(*b*) Bolts, studs, nuts, screws, keys, cotters and shafts are not usually sectioned even when the cutting plane passes through them.

(*c*) Ribs and webs are not sectioned when parallel to the cutting plane.

(*d*) The cutting plane must be indicated in the appropriate view.

(*e*) Hidden detail is not shown in sectioned views when it has already been shown in another view.

(*f*) The section shading (hatching) is normally drawn at 45° to the outline of the drawing. If the outline contains an angle of 45° then the hatching angle can be changed to avoid confusion.

(*g*) Adjacent parts are hatched in opposite directions. To show more than two adjacent parts, the spacing between the hatched lines can be varied. Some practical examples of sectioning are shown in Figs 2.36 and 2.37.

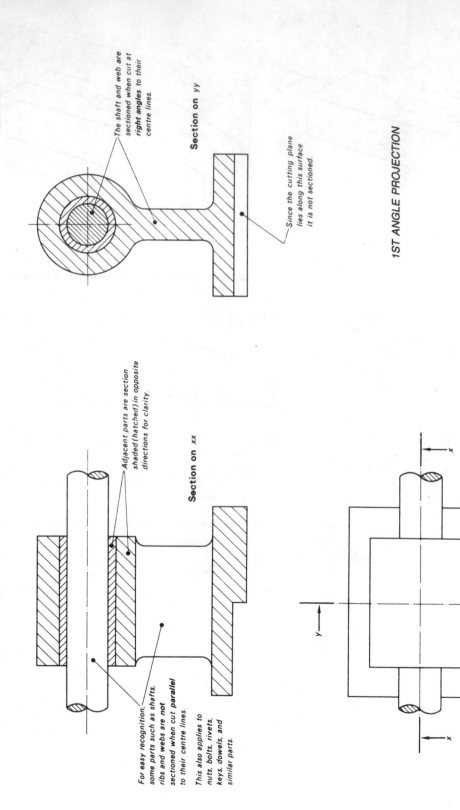

The shaft and web are sectioned when cut at right angles to their centre lines.

Section on *yy*

Since the cutting plane lies along this surface it is not sectioned.

1ST ANGLE PROJECTION

Adjacent parts are section shaded (hatched) in opposite directions for clarity.

Section on *xx*

For easy recognition, some parts such as shafts, ribs and webs are not sectioned when cut parallel to their centre lines.

This also applies to nuts, bolts, rivets, keys, dowels, and similar parts.

Fig. 2.36 Further sections

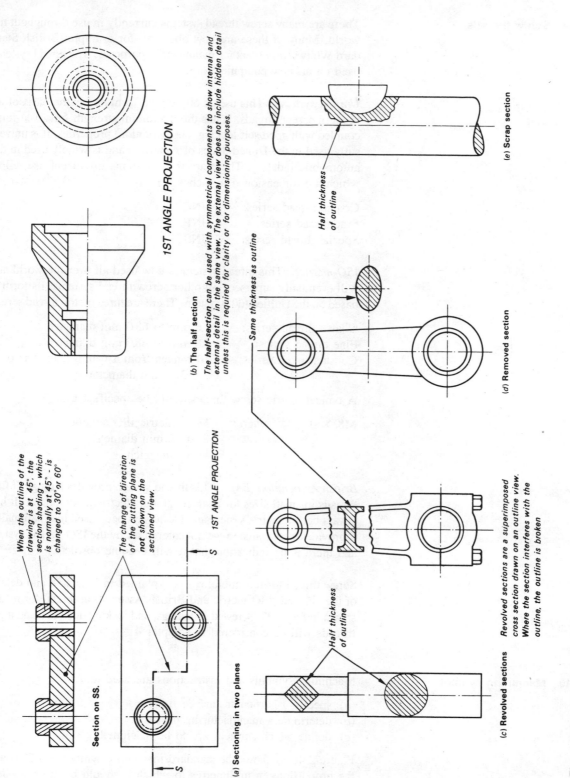

1ST ANGLE PROJECTION

(b) The half section

The half-section can be used with symmetrical components to show internal and external detail in the same view. The external view does not include hidden detail unless this is required for clarity or for dimensioning purposes.

Half thickness of outline

(e) Scrap section

Same thickness as outline

(d) Removed section

When the outline of the drawing is at 45°, the section shading - which is normally at 45° - is changed to 30° or 60°

The change of direction of the cutting plane is not shown on the sectioned view.

1ST ANGLE PROJECTION

Section on SS.

S

S

(a) Sectioning in two planes

Revolved sections are a superimposed cross section drawn on an outline view. Where the section interferes with the outline, the outline is broken

Half thickness of outline

(c) Revolved sections

Fig. 2.37 Miscellaneous sections

2.14 Screw threads

There are many screw thread systems currently in use throughout the world. Many of these are now obsolete (for example: British Standard Whitworth) except for maintenance purposes. The thread systems used on all new equipment are:

Unified system. This uses a 60° thread form and is the basis of all modern screw threads. It has the maximum strength for a vee-form coupled with good self-locking (anti-vibration) properties. It is universally used in the United States of America, and is widely used in the automobile industry. There are three systems in current use, all of which are dimensioned in inches.

Coarse thread series:	UNC
Fine thread series:	UNF
Special thread series:	UNS

ISO-metric. This thread system is now used all over the world and will eventually supersede all other screw thread systems. Its form is based on the Unified (60°) system. There are three metric thread series:

Coarse thread series:	1.6 mm to 68.0 mm diameter.
Fine thread series:	1.8 mm to 68.0 mm diameter.
Constant pitch series:	10 ranges from 1.6 mm diameter to 300 mm diameter.

A typical metric screw thread would be specified as:

M8 × 1   where: M = metric ISO system
8 = 8 mm diameter
= 1 mm pitch

British Association (BA). This thread system provides a range of fine threads in small sizes for instrument and electrical equipment. It has always been a metric system and has always been used internationally. It provides a specialist range not catered for by the ISO-metric system and there is no indication that it will become obsolete.

Screw thread form terminology is explained in Fig. 2.38 and details of the Unified, ISO-metric and British Association thread forms are shown in Fig. 2.39. Screwed fastenings, and locking devices using these threads will be considered in Chapter 4.

2.15 Machining symbols

Machining symbols and instructions are used to:

(*a*) specify a particular surface finish,
(*b*) determine a manufacturing process,
(*c*) define which surfaces are to be machined.

Figure 2.40(*a*) shows the standard machining symbol (BS 308) and the proportions in millimetres to which it should be drawn. When

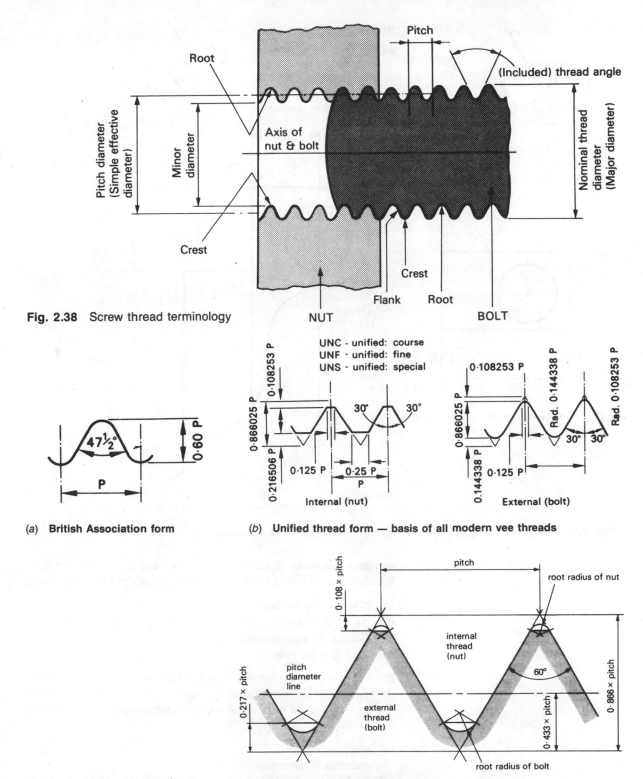

Fig. 2.38 Screw thread terminology

(a) **British Association form**

UNC - unified: course
UNF - unified: fine
UNS - unified: special

Internal (nut)

External (bolt)

(b) **Unified thread form — basis of all modern vee threads**

Fig. 2.39 Screw thread forms

(c) **ISO metric screw thread proportions**

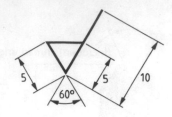

(a) Drawing a machining symbol

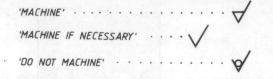

(c) The machining symbol as an instruction

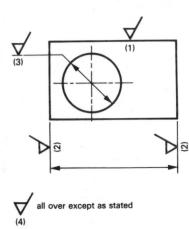

all over except as stated

(4)

(b) Applying the machining symbol.

Fig. 2.40 The machining symbol

(d) Specifying surface texture on a casting (dimensions omitted for clarity)

applied to the views of a drawing, as shown in Fig. 2.40(b), the symbol should be drawn as follows. ('Normal' means 'at right-angles to').

(a) Normal to a surface.
(b) Normal to a projection line.
(c) Normal to an extension line.
(d) As a general note.

Because a machining symbol is interpreted as a precise instruction, its form should be drawn carefully. Figure 2.40(c) shows three fundamental variations of the symbol which, in turn, instruct the craftsman to:

(a) machine,
(b) machine if necessary,
(c) do NOT machine.

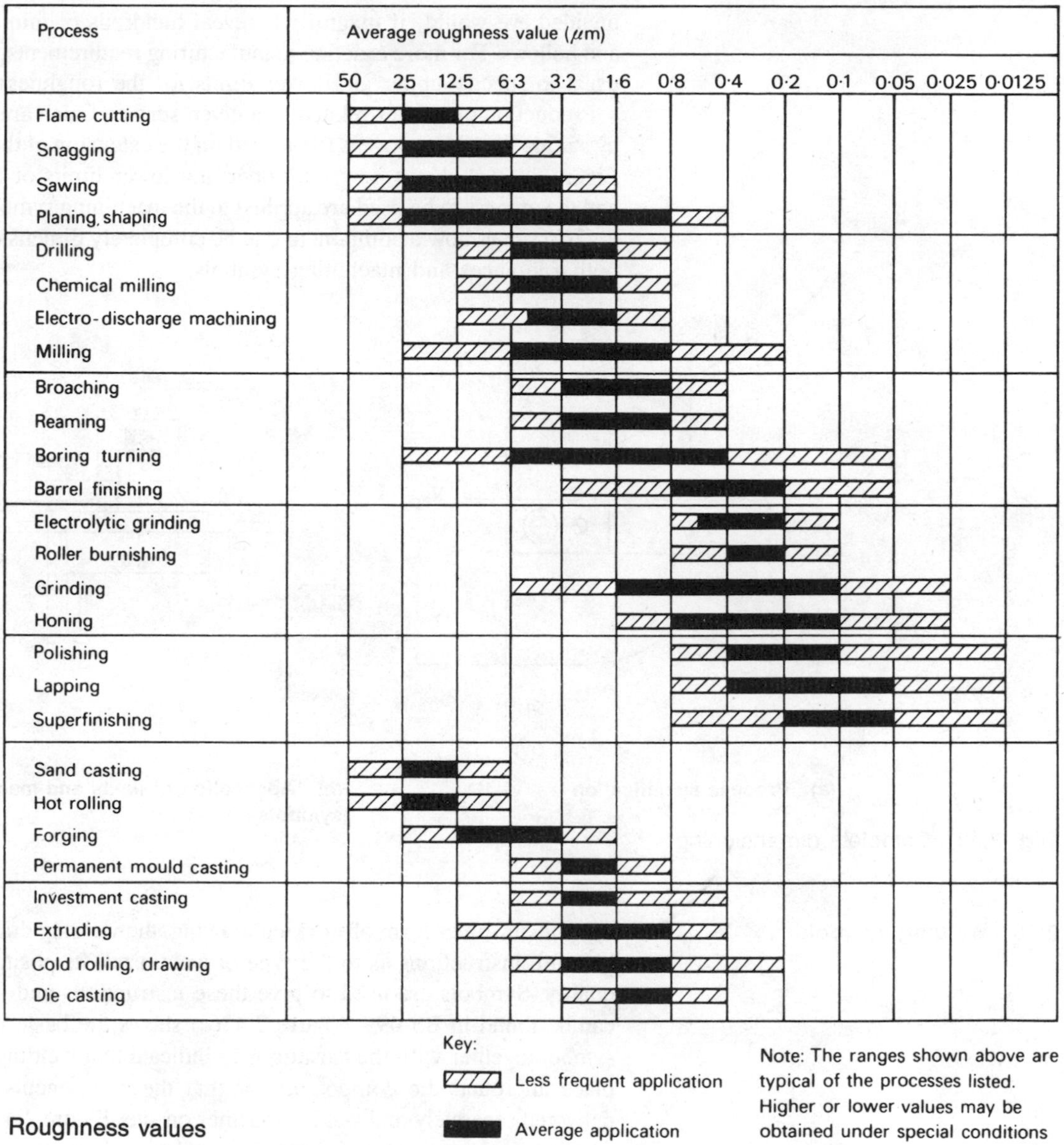

Process	\multicolumn{13}{c}{Average roughness value (μm)}

Process	50	25	12.5	6.3	3.2	1.6	0.8	0.4	0.2	0.1	0.05	0.025	0.0125
Flame cutting	▨	█											
Snagging	▨	█	▨										
Sawing	▨	█	█	▨									
Planing, shaping	▨	█	█	█	▨	▨							
Drilling				▨	█	█	▨						
Chemical milling				▨	█	█	▨						
Electro-discharge machining				▨	█	█	▨						
Milling		▨	▨	█	█	█	▨						
Broaching					▨	█	▨						
Reaming					▨	█	▨						
Boring turning			▨	▨	█	█	▨	▨					
Barrel finishing						▨	█	▨	▨				
Electrolytic grinding								▨	█	▨			
Roller burnishing								▨	█	▨			
Grinding						▨	█	█	█	▨	▨		
Honing							▨	█	█	▨	▨		
Polishing								▨	█	█	▨	▨	
Lapping								▨	▨	█	█	▨	
Superfinishing								▨	█	█	▨	▨	
Sand casting	▨	█	▨										
Hot rolling	▨	█	▨										
Forging		▨	█	█	▨								
Permanent mould casting					▨	█	▨						
Investment casting					▨	█	▨						
Extruding				▨	█	█	▨						
Cold rolling, drawing					▨	█	▨	▨					
Die casting					▨	█	▨						

Key:

▨ Less frequent application

█ Average application

Note: The ranges shown above are typical of the processes listed. Higher or lower values may be obtained under special conditions

Fig. 2.41   Roughness values

These symbols must be used carefully; one incorrect symbol or incorrect application can result in unnecessary manufacturing costs or even the scrapping of a component. Finally the symbol can be used as shown in Fig. 2.40(*d*) to indicate the quality of surface finish (acceptable roughness) and/or the production process to be used. Surface finish is usually specified in micrometres (microns) and the figure is printed above the machining symbol as shown in Fig. 2.40(*d*).

The surface finish can be related to the manufacturing process as shown in Fig. 2.41. The roughness value itself represents the average roughness of the surface. Even a surface which appears smooth to the unaided eye would, if magnified, reveal hundreds of minute humps and hollows. For more exacting manufacturing requirements, it is more usual to specify upper and lower limits for the roughness values.

Production processes related to a given surface finish are specified by stating the exact process to be used on the extension of the symbol. Figure 2.42(*a*) shows how the upper and lower limits of roughness and the process to be used are applied to the machining symbol. Figure 2.42(*b*) shows how a component can be completely dimensioned with both tolerances and machining symbols.

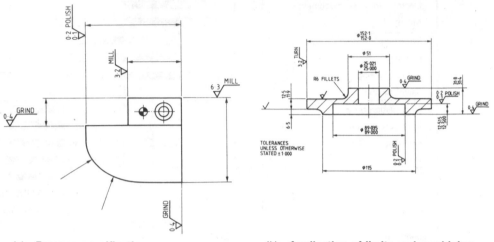

(*a*)   **Process specification**   (*b*)   **Application of limits and machining symbols**

Fig. 2.42   Complete dimensioning

2.16   Welding symbols

As with any other form of working drawing, the welding drawing must give full instructions as to the type of weld used, its position and its quality. Symbols are used to give these instructions and full details can be found in BS 499. Figure 2.43(*a*) shows the basic instruction symbol together with the variations to indicate that welding is to take place all round the component, or that the components are to be delivered separately and welded together on site. Figure 2.43(*b*) shows some typical types of weld and the symbols which represent those welds. These symbols are added to the symbol line as shown in Fig. 2.43(*c*). The examples in Fig. 2.43(*c*) show the complete conventional representation for a number of welding possibilities.

2.17   Electrical and electronic components and circuits

Just as the mechanical engineering draftsperson uses the conventions of BS 308 to represent regularly used and awkward details, so the electrical or electronic engineering draftsperson uses the symbols and conventions in BS 3939 to simplify and clarify circuit diagrams. For example Fig. 2.44(*a*) shows a pictorial representation of a simple circuit. Compare the skills required to draw this pictorial representation with the equivalent circuit diagram shown in Fig. 2.44(*b*) using cir-

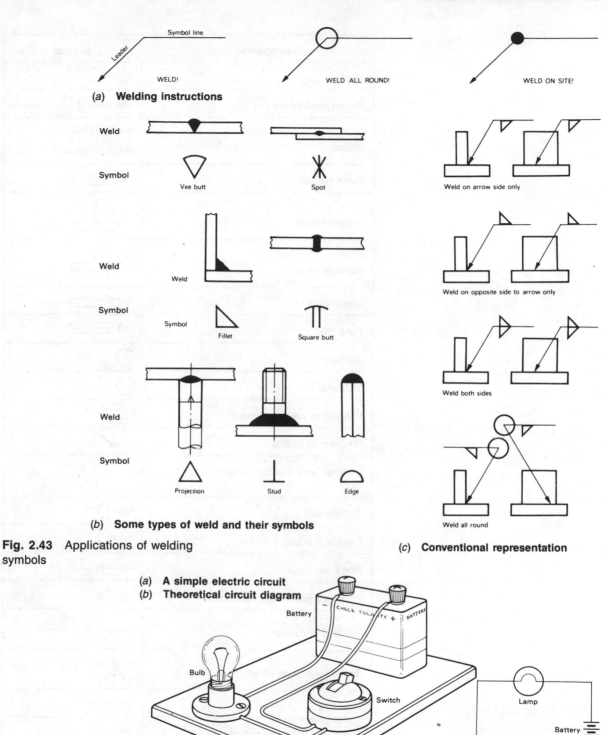

Symbol line

Leader

WELD!

WELD ALL ROUND!

WELD ON SITE!

(a) **Welding instructions**

Weld

Symbol

Vee butt

Spot

Weld on arrow side only

Weld

Weld

Symbol

Symbol

Fillet

Square butt

Weld on opposite side to arrow only

Weld

Symbol

Projection

Stud

Edge

Weld both sides

(b) **Some types of weld and their symbols**

Weld all round

Fig. 2.43 Applications of welding symbols

(c) **Conventional representation**

(a) **A simple electric circuit**
(b) **Theoretical circuit diagram**

Battery

CHECK POLARITY +   BATTERY

Bulb

Switch

Lamp

Battery

Switch

Fig. 2.44 The circuit diagram

Description	Symbol
Earth	
Primary/secondary cell	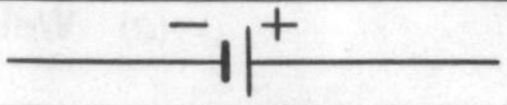
Battery	
Fixed resistor	
Potentiometer	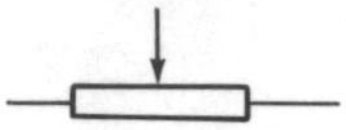
Variable resistor	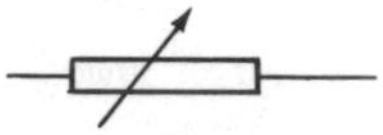
Detachable link	
Fuse link	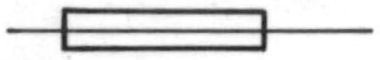
Ammeter	
Voltmeter	
Integrating watt-hour meter	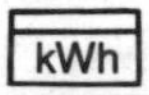 kWh
Single-pole switch	
General lamp symbol including signal lamp	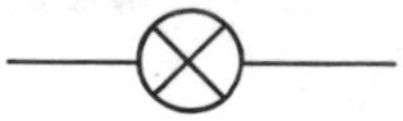
Electric bell	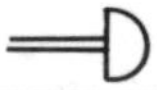
Electric buzzer	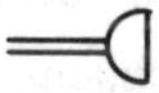
Fixed capacitor	
Polarised (electrolytic) capacitor	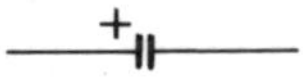
Variable capacitor	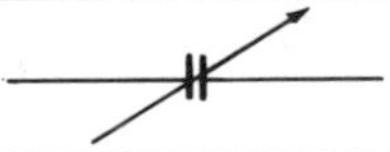
Transformer (iron cored)	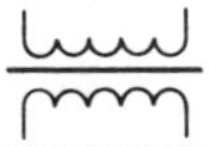
Solenoid	
Iron-cored inductor (choke)	
P—N diode	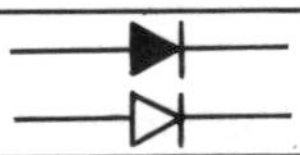

Fig. 2.45 Theoretical circuit symbols

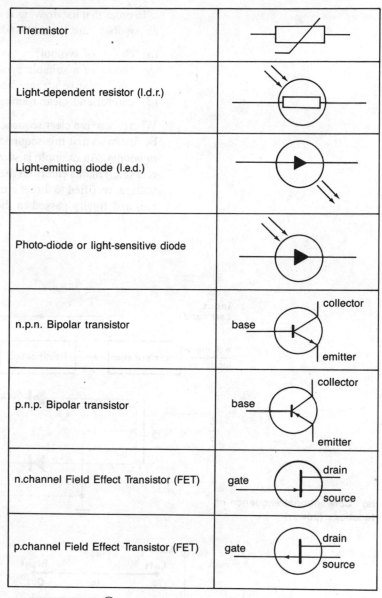

Thermistor	
Light-dependent resistor (l.d.r.)	
Light-emitting diode (l.e.d.)	
Photo-diode or light-sensitive diode	
n.p.n. Bipolar transistor	base, collector, emitter
p.n.p. Bipolar transistor	base, collector, emitter
n.channel Field Effect Transistor (FET)	gate, drain, source
p.channel Field Effect Transistor (FET)	gate, drain, source

Note: the envelope ◯ may be omitted from these symbols providing it is not connected to one of the electrodes or to earth and providing its omission does not lead to confusion.

Fig. 2.45
continued

cuit symbols selected from BS 3939. It is only possible, within the limits of this chapter, to consider a few of the symbols and conventions of this very extensive standard which is published in 13 sections and covers every application and component ranging from power generation and distribution, through telecommunications to the latest state-of-the-art, computer logic circuits. However a selection of symbols are shown in Fig. 2.45.

In order that the drawing is as informative and free from ambiguities as possible, care should be taken over its layout by paying attention to:

(*a*) choice of symbol;
(*b*) choice of a suitable symbol orientation;
(*c*) good relative arrangement of symbols;
(*d*) careful and clear routing of interconnections.

Where there is a clear sequence from cause to effect the diagram should be drawn so that this sequence is shown from *left* to *right* and/or *top to bottom*. An example is shown in Fig. 2.46(*a*). The alternating current (a.c.) mains enters on the left, is transformed to a higher or lower voltage, rectified to direct current (d.c.), smoothed to remove the ripple, and finally passed to the load which is shown on the right.

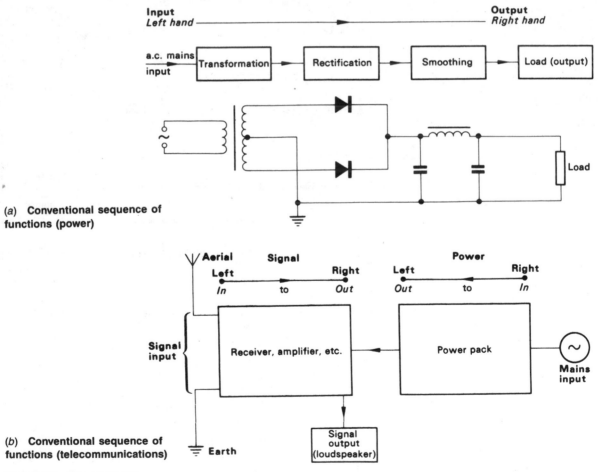

(*a*) **Conventional sequence of functions (power)**

(*b*) **Conventional sequence of functions (telecommunications)**

Fig. 2.46 Circuit layout

The exception to this is in telecommunication circuits (radios, televisions, amplifiers, etc.) where the signal takes priority and flows through the diagram from left to right, and the power supply flows from right to left as shown in Fig. 2.46(*b*).

The following general points should also be observed.

(*a*) It is normal to show all circuit diagrams with the switches or relays in the *open* position. That is, with the circuit in its unoperated, no current condition with all supplies disconnected. For example in Fig. 2.44(*b*) the switch is shown in the open or 'off' position.

(*b*) If multi-switching devices are shown, the contacts must be in mutually consistent positions irrespective of whether the circuit is inoperative or not as shown in Fig. 2.47. In this circuit the relay holds S_2 *open* when the circuit is energised. When the circuit is switched off or the supply fails S_2 closes and the alarm rings. As drawn, the switch S_1 is shown correctly in the open position. Since this is equivalent to a supply failure, S_2 is shown in the *closed* position and the alarm will sound. That is, the position of S_2 is consistent with that of S_1.

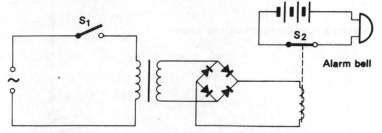

Fig. 2.47 Mutually consistent switching

The switch S_1 is shown correctly in the open position.
Since this is equivalent to a supply failure, S_2 is shown correctly
in the **closed** position and the alarm will sound.

(*c*) Explanatory notes should be used where the meaning cannot be expressed conveniently in any other way as, for example, the ohmic value and power rating of a resistor. Abbreviations should follow BS 1991 for 'Letter Symbols, Signs and Abbreviations'.

(*d*) Lines representing conductors should, in general, be horizontal or vertical. However, it is permissible and sometimes preferable to use oblique lines. For example when drawing a *bridge circuit* or a *lattice network* as shown in Fig. 2.48(*a*). Conductors crossing *without connecting* are represented by the symbol shown in Fig. 2.48(*b*) where the conductors are in contact they should be drawn as shown in Fig. 2.48(*c*) and 2.48(*d*). The 'dot' at the termination indicates the connection. (Imagine it to be the blob of solder making the connection.) Where it is necessary to draw the vertical conductors 'in line' the junctions should be as shown in Fig. 2.48(*e*).

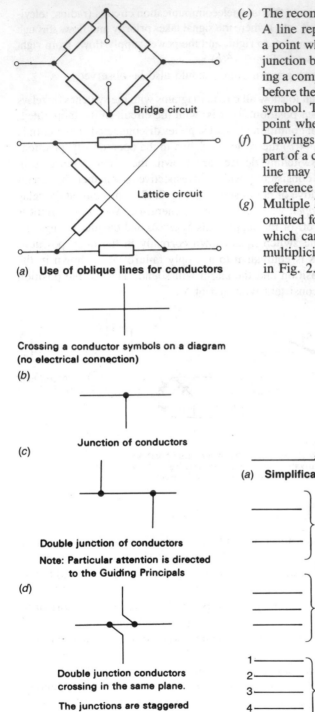

(a) **Use of oblique lines for conductors**

Crossing a conductor symbols on a diagram
(no electrical connection)

(b)

(c) **Junction of conductors**

Double junction of conductors

**Note: Particular attention is directed
to the Guiding Principals**

(d)

**Double junction conductors
crossing in the same plane.**

**The junctions are staggered
as shown (Note D.5 BS 3939)**

(e)

Fig. 2.48 Crossing and connection
of conductors

(e) The recommendations for lines changing direction are as follows. A line representing a conductor should not change direction at a point where it crosses another line, nor should it cross over a junction between other lines. Lines representing conductors leaving a component symbol should be continued for a short distance before they change direction, cross another line, or meet another symbol. Thus lines representing conductors must not meet at the point where they enter or leave a symbol.

(f) Drawings may be simplified by omitting lines where they cross part of a circuit diagram. Under these circumstances most of the line may be omitted and the connection indicated by a number reference as shown in Fig. 2.49(a).

(g) Multiple lines representing conductors that are parallel may be omitted for most of their length, leaving short lines at each end which can be grouped and cross-referenced. Alternatively, the multiplicity of lines can be represented by a single line as shown in Fig. 2.49(b).

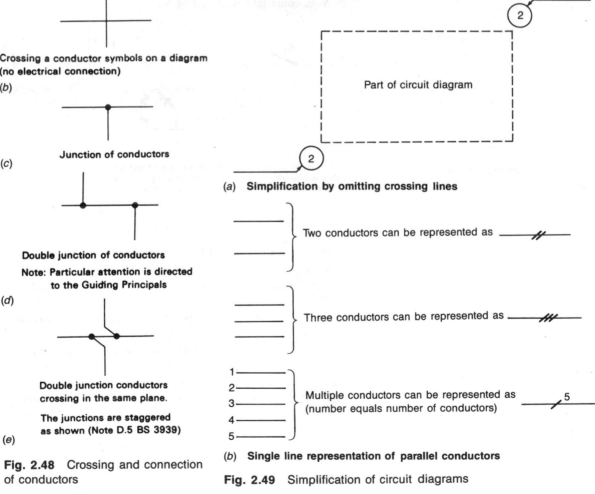

(a) **Simplification by omitting crossing lines**

(b) **Single line representation of parallel conductors**

Fig. 2.49 Simplification of circuit diagrams

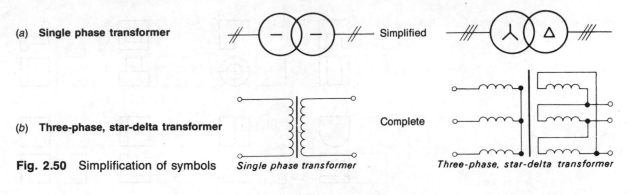

(a) **Single phase transformer** — Simplified

(b) **Three-phase, star-delta transformer** — Complete

Fig. 2.50 Simplification of symbols   *Single phase transformer*   *Three-phase, star-delta transformer*

(h) Simplification of symbols. BS 3939 allows for the simplification of symbols where their use will not cause confusion. For example Fig. 2.50 shows how simplified symbols may be applied to single-phase and three-phase transformers.

Some typical circuit diagrams making use of the above symbols and conventions are shown in Chapters 6 and 7.

Problems

Section A

1   Figure 2.51 shows two views of twelve simple objects in *first angle* projection. Trace the views and add the third, missing view in the boxes provided.

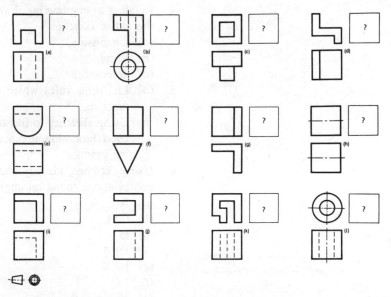

Fig. 2.51

2   Figure 2.52 shows two views of twelve simple objects in *third angle* projection. Trace the views and add the third, missing view in the boxes provided.

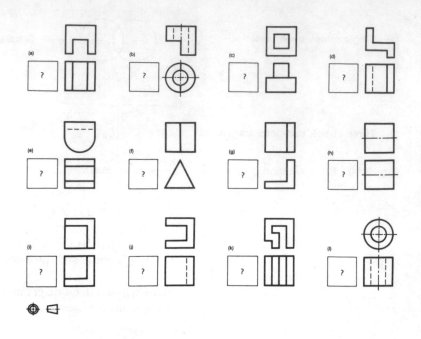

Fig. 2.52

3 By means of sketches of simple objects of your own choice, show what is meant by:
   (a) isometric view;
   (b) oblique view;
   (c) auxiliary view.
4 Relating to engineering drawings, explain the following terms with the aid of sketches.
   (a) Dimension.
   (b) Limit.
   (c) Tolerance.
5 (a) List *three* rules which provide a basis for good dimensioning practice.
   (b) Using sketches to illustrate your answer, give five examples of *good* dimensioning practice and five examples of *bad* dimensioning practice.
6 Using sketches, identify and describe the meaning of the following abbreviations found on engineering drawings.
   (a) A/F
   (b) CYL
   (c) Ø
   (d) DIM
   (e) HEX
   (f) LG
   (g) NO
   (h) R
   (i) SCR
   (j) □

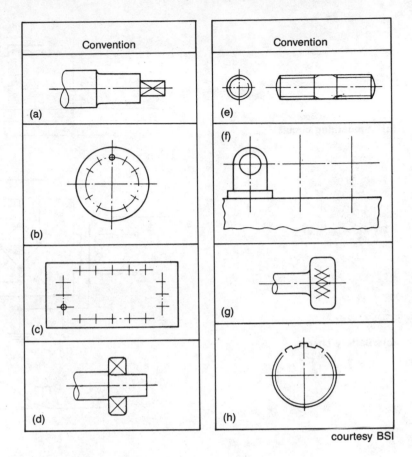

Fig. 2.53

courtesy BSI

7   Figure 2.53 shows a number of conventions drawn to BS 308. Describe what the conventions stand for and under what circumstances they would be used.

8   Figure 2.54 shows some electrical circuits. Redraw the circuits and name the symbols shown.

9   Figure 2.55 shows some welding symbols (BS 499). Sketch the joint to which each symbol refers.

10   Figure 2.56 shows some variations of the machining symbol. Describe the meaning of each symbol.

Section B

11   Figure 2.57 shows a pictorial view of a die support block. Draw full size in first OR third angle projection:
     (*a*) an end elevation in the direction of arrow A;
     (*b*) an elevation in the direction of arrow B;
     (*c*) a section plan view on YY.

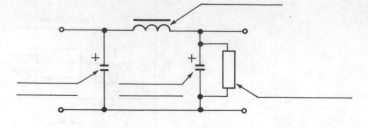

(a) **Smoothing circuit**

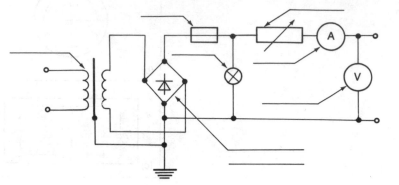

(b) **Battery charger**

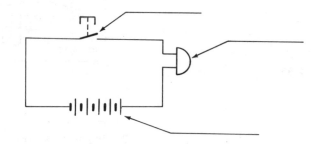

(c) **Alarm circuit**

Fig. 2.54

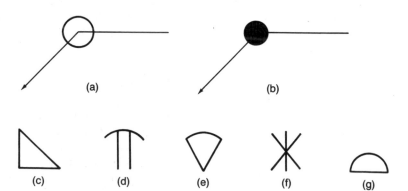

Fig. 2.55

(a)

(b)

(c)

(d)

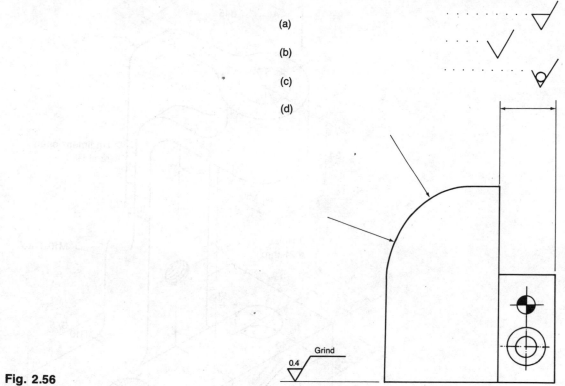

0.4 Grind

Fig. 2.56

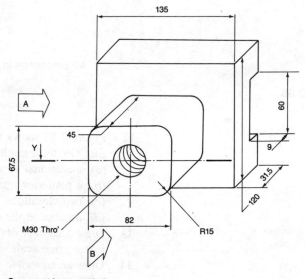

A

Y

45

67.5

82

M30 Thro'

R15

B

135

60

9

31.5

120

Component is symmetrical
about ₵'s
All dimensions in
millimetres

Fig. 2.57

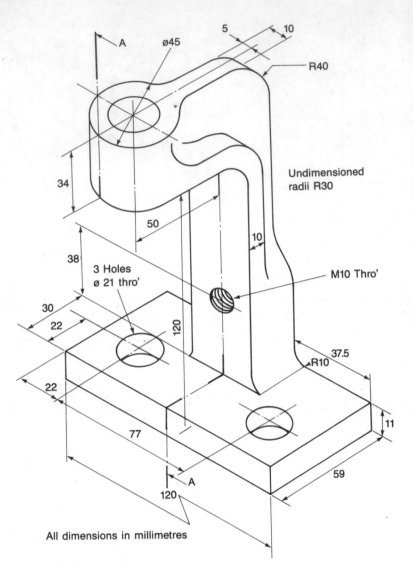

Fig. 2.58

12 Figure 2.58 shows a bracket. Draw the bracket to a scale of 1:1. in first OR third angle projection:
    (*a*) a sectional view on AA;
    (*b*) a plan view (projected from section AA);
    (*c*) two elevations (projected either side of section AA);
    (*d*) dimension the drawing.

13 Draw a *cabinet oblique* view of the component shown in Fig. 2.59 to the same scale.

14 Draw an *isometric* view of the component shown in Fig. 2.59 to the same scale.

15 Figure 2.58 shows a bracket. Draw the bracket to a scale of 1:1 in the items are welded together.

View direction of
arrow X

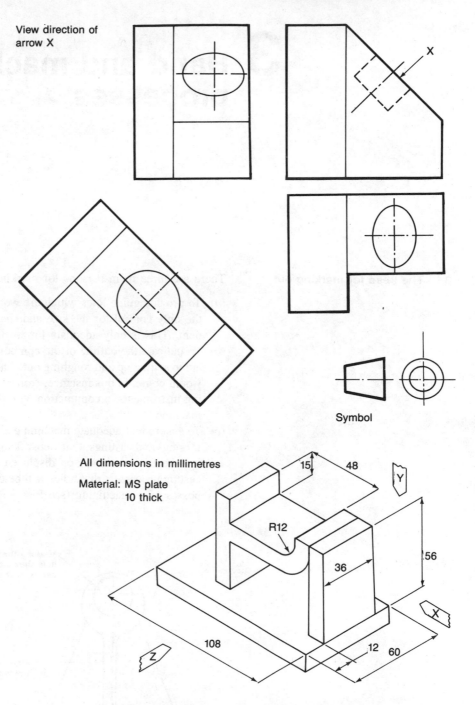

Fig. 2.59

Symbol

All dimensions in millimetres

Material: MS plate
10 thick

15   48

R12

36

56

108

12   60

Fig. 2.60

(*a*) Draw, full size, in first angle projection, a plan view in the direction of arrow 'Y' and elevations in the directions of arrows 'X' and 'Z'.

(*b*) Complete the drawing by adding the appropriate welding symbols and any other information necessary for manufacture.

3 Hand and machine processes

3.1 The need for marking out

There are three main reasons for marking out.

(a) To provide guide lines which are worked to, and which provide the only control for the size and shape of the finished component. This is only adequate for work of low accuracy.

(b) To indicate the outline of the component to the machinist as an aid to setting up and roughing out. The final dimensional control would come, in this instance, from the use of precision measuring instruments in conjunction with the micrometer dials on the machine.

(c) To ensure that adequate machining allowances have been left on forgings and castings; that webs, flanges and cores have not been incorrectly positioned or displaced during the moulding and casting processes; that holes will be centrally positioned in their bosses after machining (see Fig. 3.1).

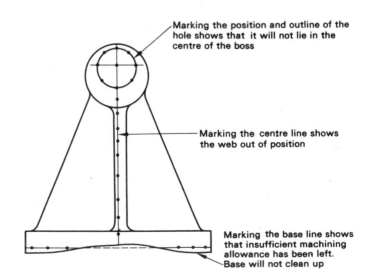

Marking the position and outline of the hole shows that it will not lie in the centre of the boss

Marking the centre line shows the web out of position

Marking the base line shows that insufficient machining allowance has been left. Base will not clean up

Fig. 3.1 Checking a casting

3.2 The scribed line

For accurate work a clean, fine line is required. This is produced using a scriber using a sharp point guided by a straight edge or template.

For scribing circular lines up to about 150 mm diameter, *dividers* are used. Above this diameter *trammels* are used. Figure 3.2 shows examples of these instruments. In soft material the pivot point can be pressed into the surface being marked. In harder materials a small indentation is made with a 'dot' punch to locate the pivot point. This is similar to a centre punch (used to start drills) but is finer and has a more acute point. As well as being used to scribe circles, dividers and trammels are also used for stepping off distances as shown in Fig. 3.2.

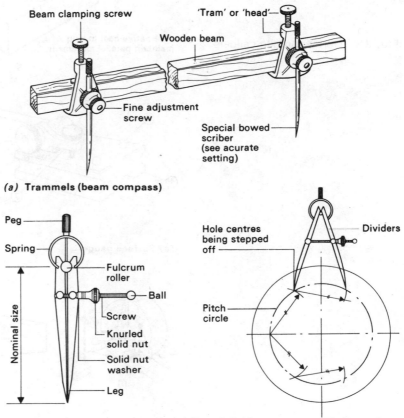

(a) Trammels (beam compass)

(b) Dividers

(c) Use of dividers to step off hole centres

Fig. 3.2 Dividers and trammels

Hermaphrodite calipers (also known as 'odd-leg' and 'jenny' calipers) are used for scribing lines which are parallel to an edge or surface throughout their lengths. Examples of such calipers are shown in Fig. 3.3.

A *surface gauge* or *scribing block* is used for scribing lines and setting surfaces parallel to a datum surface. It can also be used for checking parallelism and setting components in a machine. A typical instrument is shown in Fig. 3.4(a) and applications are shown in Fig. 3.4(b).

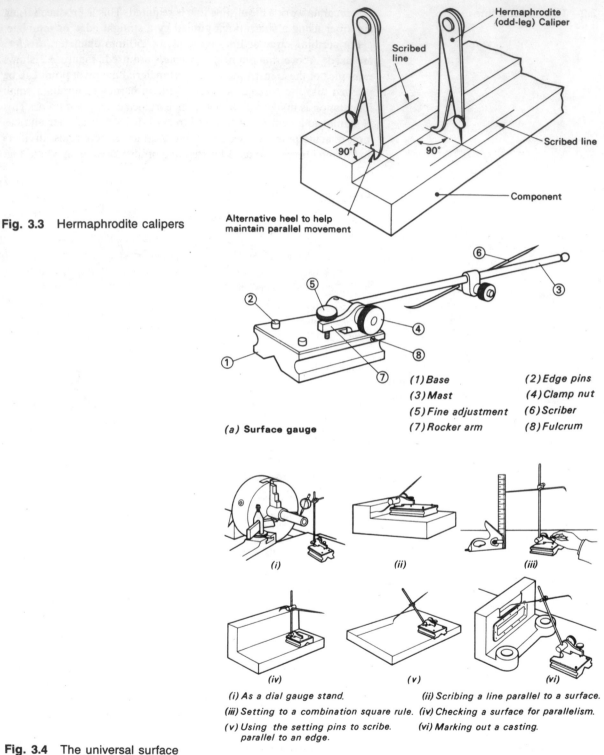

Fig. 3.3 Hermaphrodite calipers

(1) Base   (2) Edge pins
(3) Mast   (4) Clamp nut
(5) Fine adjustment   (6) Scriber
(7) Rocker arm   (8) Fulcrum

(a) **Surface gauge**

(i) As a dial gauge stand.   (ii) Scribing a line parallel to a surface.
(iii) Setting to a combination square rule.   (iv) Checking a surface for parallelism.
(v) Using the setting pins to scribe.   (vi) Marking out a casting.
     parallel to an edge.

Fig. 3.4 The universal surface
gauge

(b) **Typical applications**

(Courtesy of Moore & Wright Ltd)

To preserve the scribed line, it is usual to make a series of dots along it with a dot punch in order that:

(a) the line may be restored if it becomes obliterated;

(b) the dots may act as a 'witness' after cutting out or machining the component, that is, machining or filing to the line will leave half the dot holes visible as proof that the 'line has been split'. This is important as the line will no longer be visible.

3.3 Datum lines and datum edges

A *datum* can be defined as a fixed point, line, edge or surface from which a measurement can be taken. The distance from a datum to a hole centre or other feature is called an *ordinate*. In practice two such dimensions are required to fix the position of a point on a flat surface. These two ordinates are known as *coordinates*. There are two systems of coordinates in common use.

(a) *Rectangular coordinates*: The point is positioned by a pair of ordinates (coordinates) lying at right angles to each other and at right angles to the axes or datum edges from which they are measured. This system requires the preparation of two mutually perpendicular datum edges before marking out can commence. Figure 3.5(a) shows an example of the centre of a hole dimensioned by means of rectangular coordinates.

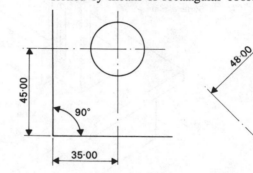

(a) **Rectangular coordinates**   (b) **Polar coordinates**

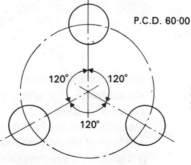 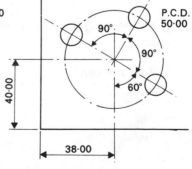

(c) **Polar coordinates applied to holes on a pitch circle**
(Note: P.C.D. = Pitch circle diameter)

(d) **Combined coordinates**
Dimensions in mm

Fig. 3.5 Coordinates

(b) *Polar coordinates*: In this instance the coordinates consist of a linear distance and an angle. Dimensioning by this technique is useful when work is to be machined with the aid of a rotary table. Figure 3.5(b) shows the principle of dimensioning a hole centre using polar coordinates, whilst Fig. 3.5(c) shows how polar coordinates can be applied to the dimensioning of holes on a pitch circle. It can be seen that polar coordinates use a point datum instead of edge datums. In practice, this point datum would have to be established from edge datums by means of rectangular coordinates as shown in Fig. 3.5(d), thus becoming a *secondary* datum.

Figure 3.6 shows a typical marking out set up using a surface table as a datum.

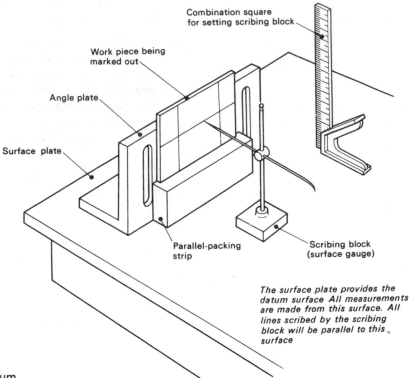

Combination square for setting scribing block

Work piece being marked out

Angle plate

Surface plate

Parallel-packing strip

Scribing block (surface gauge)

The surface plate provides the datum surface All measurements are made from this surface. All lines scribed by the scribing block will be parallel to this surface

Fig. 3.6 Marking out from a datum surface

3.4 Measurement

Accurate measurement is the basis of good engineering practice. Measurement is a comparative process, whereby the object being measured is compared with a known standard. Table 3.1 lists the more generally used workshop measuring devices which become the local standards of linear measurement with which the object being measured is compared. These workshop standards of length are themselves based upon the national and international standards of length. Thus the components produced in one factory will be interchangeable with similar components produced in another factory.

Table 3.1 Workshop standards of length

Name	Range (in mm)	Reading accuracy
Steel rule	150 to 1000	0.5 mm
Vernier caliper	0/150 to 0/2000	0.02 mm
Micrometer caliper	0/25 to 1800	0.01 mm
Slip gauges	1.025 to 327 (105-piece set)	0.0025 mm

3.5 The steel rule

The steel rule is frequently used in the workshop for measuring components of limited accuracy because of the difficulty of sighting the graduations in line with the feature being measured. Ways of minimising these errors when measuring with a steel rule are shown in Fig. 3.7.

A good rule should be looked after carefully to prevent damage to its scales and edges. It should never be used as a scraper or a screwdriver, and it should never be used to remove swarf from machine tool worktable 'Tee' slots. After use the rule should be wiped clean and lightly oiled to prevent rusting as dulling of the surface will make it difficult to read.

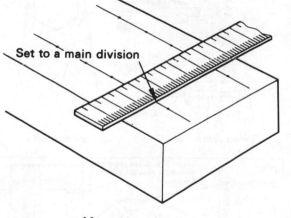

Measuring the distance between two scribed lines

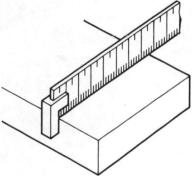

Measuring the distance between two faces using a hook rule

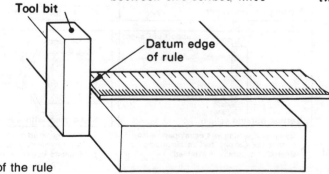

Measuring the distance between two faces using a steel rule and abutment

Fig. 3.7 Use of the rule

3.6 Calipers

Calipers are used in conjunction with a rule so as to transfer the distance across or between the faces of a component in such a way as to reduce sighting errors. Firm-joint calipers are usually used in the larger sizes. Spring joint calipers are usually used for fine work, as in instrument-making and tool-making. Examples of internal and external calipers, both firm joint and spring joint, together with examples of their uses are shown in Fig. 3.8.

The accurate use of calipers depends upon a highly developed sense of feel that can only be acquired by practice. When using calipers, the following rules should be observed.

(*a*) Hold the caliper gently and near the joint.

(*b*) Hold it square to the work.

(*c*) No force should be used to 'spring' it over the work. Contact should only just be felt.

(*d*) The caliper should be handled and laid down gently to avoid disturbing the setting.

(*e*) Lathe work should be *stationary* when taking measurements. This is essential for *safety*, and *accuracy*.

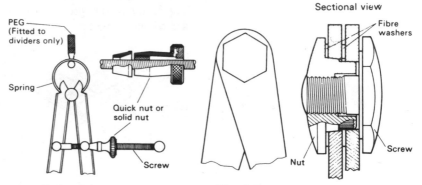

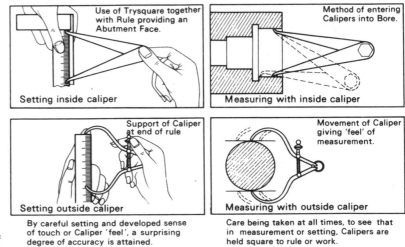

Fig. 3.8 Construction and use of calipers

By careful setting and developed sense of touch or Caliper 'feel', a surprising degree of accuracy is attained.

Care being taken at all times, to see that in measurement or setting, Calipers are held square to rule or work.

(Courtesy of Moore & Wright Ltd)

3.7 The micrometer caliper

Most engineering work has to be measured to a much greater accuracy than is possible with a rule, even when aided by the use of calipers. To achieve this greater precision, measuring equipment of greater accuracy and sensitivity must be used.

One of the most familiar precision measuring instruments found in the workshop is the *micrometer caliper*. The construction of a micrometer caliper and details of its more important parts and features are shown in Fig. 3.9. The operation of the micrometer depends upon the principle that the distance moved by a nut along a screw is proportional to the number of revolutions made by the nut. Therefore by controlling the number of revolutions and fractions of a revolution made by the nut, the distance it moves along the screw can be accurately controlled. This principle forms the basis of the micrometer as applied to measuring devices and the movement of machine tool slideways. The movement of the screw and nut is relative. The same arguments apply if the nut remains stationary and the screw is rotated.

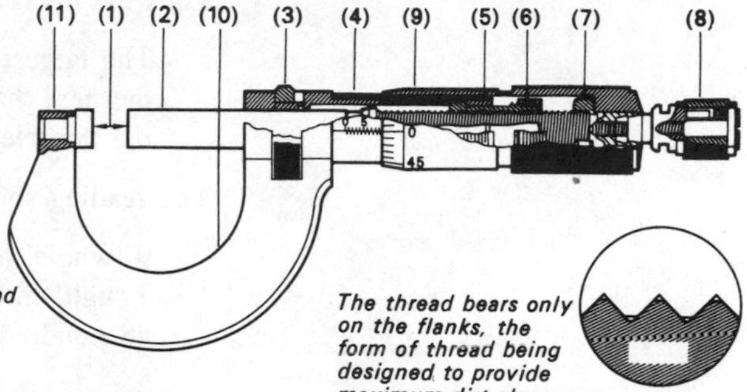

Pearl chrome plated to eliminate glare and give easy reading

The thread bears only on the flanks, the form of thread being designed to provide maximum dirt clearance and adjustment

(Courtesy of Moore & Wright Ltd.)

(1) **Spindle and anvil faces** — Glass hard and optically flat, also available with **Tungsten carbide** faces
(2) **Spindle** — Thread ground, and made from alloy steel, hardened throughout, and stabilised
(3) **Locknut** — effective at any position. Spindle retained in perfect alignment
(4) **Barrel** — Adjustable for zero setting. Accurately divided and clearly marked. Pearl chrome plated
(5) **Main nut** — Length of thread ensures long working life
(6) **Screw adjusting nut** — For effective adjustment of main nut
(7) **Thimble adjusting nut** — Controls position of thimble
(8) **Ratchet** — Ensures a constant measuring pressure
(9) **Thimble** — Accurately divided and every graduation clearly numbered
(10) **Steel frame** — Drop forged. Marked with useful decimal equivalents
(11) **Anvil end** — Cutaway frame facilitates usage in narrow slots

Fig. 3.9 Construction of the micrometer caliper

These principles are incorporated in the micrometer. The screw thread is rotated by the thimble which has a scale to indicate the 'partial' revolutions. The barrel of the instrument has a scale which indicates the 'whole' revolutions. In a standard metric micrometer the screw has a lead of 0.5 millimetre and the thimble and barrel are graduated as shown in Fig. 3.10.

Since the lead of the screw of a standard metric micrometer is 0.5 millimetre and the barrel divisions are 0.5 millimetre apart, one revolution of the thimble and screw moves the thimble along the barrel by

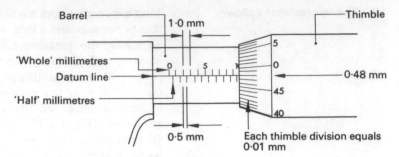

Fig. 3.10 Micrometer scales (metric)

one barrel division (0.5 mm). The barrel divisions are placed on alternate sides of the datum line for clarity. Further, since the thimble has 50 divisions and one revolution of the thimble equals 0.5 millimetre, then a movement of *one thimble division* equals:

0.5 millimetre/50 divisions = 0.01 millimetre

Thus the micrometer reading is given by:

The largest visible 'whole' millimetre graduation +
the next 'half' millimetre graduation, if visible +
the thimble division coincident with the datum line.

The reading shown in Fig. 3.10 is as follows:

9 'whole' millimetres	= 9.00
1 'half' millimetre	= 0.50
48 hundredths of a millimetre	= 0.48
	= 9.98 mm

Unless a micrometer is properly looked after it will soon lose its high initial accuracy. To maintain this accuracy the following precautions should be observed:

(*a*) wipe the anvils and work to be measured perfectly clean before making a measurement;

(*b*) do not use excessive measuring pressure (two 'clicks' of the ratchet is sufficient);

(*c*) do not leave the anvil faces in contact when not in use;

(*d*) stop the machine before measuring. Attempting to take measurements with the machine working can lead to very serious accidents. This rule applies to all measuring instruments.

From time to time the micrometer should be checked and adjusted using the double ended spanner to be found in its case. The procedure for doing this is as follows.

(*a*) Any looseness in the screw can be taken up by a slight turn of the screw adjusting nut — item 6, Fig. 3.9.

(*b*) Zero error. Periodically the anvil faces should be cleaned and closed using the ratchet to give the correct measuring pressure. If the zero line of the thimble does not coincide with the datum line on the barrel, turn the barrel — item 4, Fig. 3.9 — in the frame with the 'C' end of the spanner until the datum line coincides with the zero.

The micrometer principle can also be applied to the *internal micrometer* (Fig. 3.11(*a*)) which is used for measuring bores from 50 millimetres diameter to 210 millimetres diameter; the micrometer *cylinder gauge* (Fig. 3.11(*b*)) which can be used for smaller diameter holes, and the *depth micrometer* (Fig. 3.11(*c*)). Care must be taken when using the micrometer depth gauge since its scales are reversed.

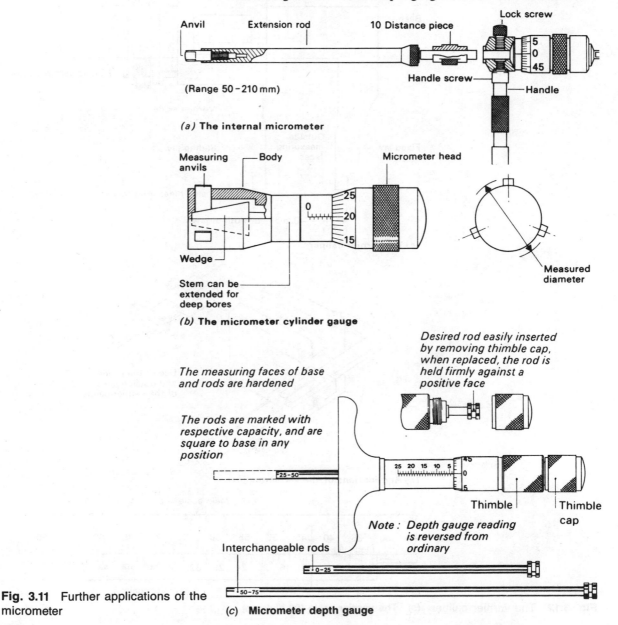

(a) **The internal micrometer**

(b) **The micrometer cylinder gauge**

Fig. 3.11 Further applications of the micrometer

(c) **Micrometer depth gauge**

3.8 The vernier caliper

Figure 3.12 shows a vernier caliper. It can be seen that, unlike the micrometer caliper, the vernier caliper can make inside and outside measurements with the one instrument. Furthermore, the vernier caliper reads from zero over the full length of its beam scale, whereas the micrometer can only read over a range of 25 millimetres. (In other words, different micrometers are required for 0−25 mm, 25−50 mm, 50−75 mm, and so on.)

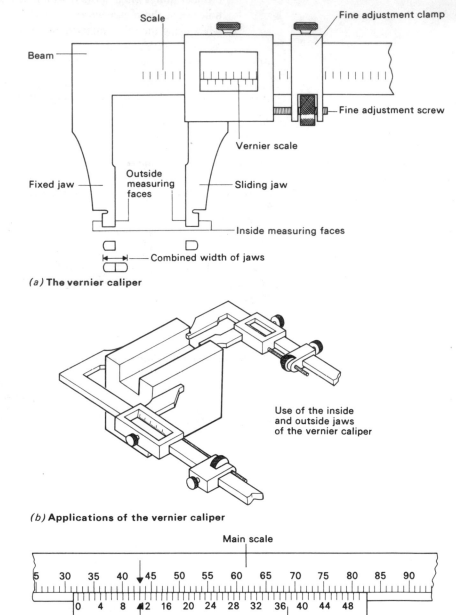

(a) The vernier caliper

(b) Applications of the vernier caliper

Fig. 3.12 The vernier caliper (c) **The vernier scale (50 division)**

Note: For inside readings, the thickness of the jaws has to be added to the scale reading (Figure 3.12(*b*)).

Unfortunately the vernier caliper does not give such accurate readings as the micrometer caliper for the following reasons.

(*a*) It is difficult to obtain a correct 'feel' due its size and weight and the lack of a 'ratchet' to control the contact pressure with the work.
(*b*) The scales are difficult to read even with the aid of a magnifying glass.
(*c*) The reading accuracy of the scales is only 0.02 millimetre (cf. micrometer: 0.01 millimetre).

Figure 3.12(*c*) shows a typical 50-division vernier scale as used on modern metric measuring instruments. The main scale is marked in 'whole' millimetres. The vernier scale is divided into 50 units which occupy 49 millimetres on the main scale. Thus each unit division of the vernier scale shows a difference of 1/50 millimetre (0.02 mm). To read the vernier:

(*a*) note the position of the 'zero' mark on the vernier scale and the number of 'whole' millimetres to the *left* of the zero mark.
(*b*) note the vernier reading at the point where the vernier and main scale graduations coincide. The additional, fractional reading will be equal to this vernier reading multiplied by 0.02 millimetre.

Thus the reading shown in Fig. 3.12(*c*) is:

$$
\begin{array}{lll}
\text{32 'whole' millimetres} & = & 32.00 \\
\text{plus } 11 \times 0.02 \text{ millimetres} & = & \underline{0.22} \\
 & & \underline{32.22} \text{ mm}
\end{array}
$$

The vernier principle is also applied to the vernier height gauge which is shown in Fig. 3.13 together with applications. As with all measuring instruments, vernier calipers and height gauges should be carefully cleaned before and after use and kept in the case provided. They should not be dropped, nor should excessive force be used when making a measurement.

3.9 The measurement of angles

Figure 3.14(*a*) shows a typical engineer's try-square which is used for marking out lines at right angles (90°) to an edge, or for checking that surfaces are at right angles to each other. Two surfaces or lines at right angles to each other are said to be:

(*a*) perpendicular to each other,
(*b*) mutually perpendicular,
(*c*) 'square' to each other.

All of which mean the same thing.

Figure 3.14(*b*) shows two applications of the try-square. In the first example the stock is placed against the edge of the work AB and slid

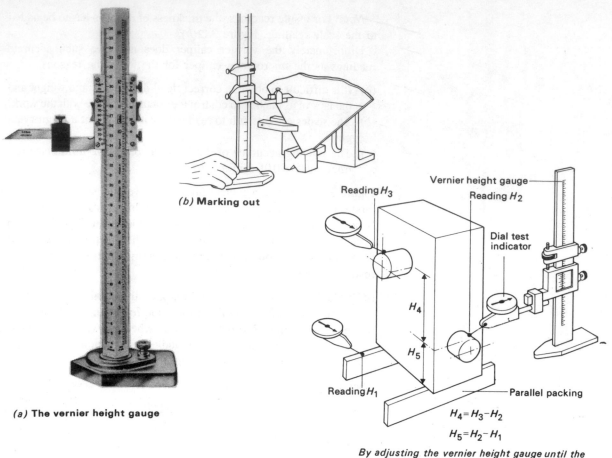

(b) Marking out

Reading H_3

Vernier height gauge

Reading H_2

Dial test indicator

H_4

H_5

Reading H_1

Parallel packing

$H_4 = H_3 - H_2$

$H_5 = H_2 - H_1$

By adjusting the vernier height gauge until the dial test indicator reads zero for each measurement taken, errors of 'feel' are removed

(a) The vernier height gauge

Fig. 3.13 The vernier height gauge

(c) Measuring the height of a surface

gently downwards until the blade comes into contact with the edge BC. Any lack of squareness will allow light to be seen between the edge BC and the try-square blade.

It is not always convenient to hold a large component and try-square up to the light. The second example shows an alternative method using a surface plate as a datum surface. The squareness of the component face iş checked using feeler gauges as shown. If the face is square to the base then the gap between it and the try-square blade will be constant.

Try-squares are precision instruments and they should be treated with care if they are to retain their initial accuracy. They should be kept clean and not dropped. They should be lightly oiled after use. They should be kept away from other bench tools to avoid burrs being knocked up on the edges of the blade or the stock. They should be checked for squareness at regular intervals.

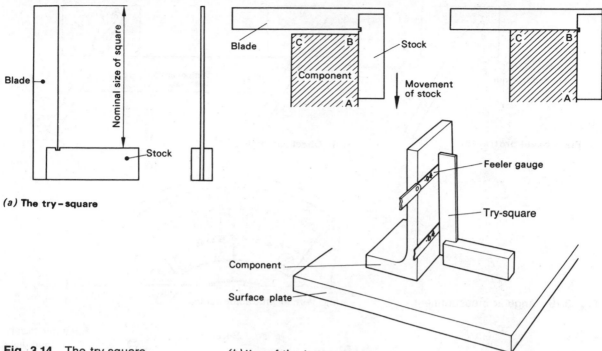

(a) **The try-square**

Fig. 3.14   The try-square

(b) **Uses of the trysquare**

Figure 3.15 shows a simple protractor and how it is used. It is used for measuring angles other than 90°, but it has only limited accuracy, (±0.5°). The simple protractor can be extended in accuracy by the addition of a circular vernier scale to improve its reading accuracy as shown in Fig. 3.15(c). The main scale is graduated in degrees of arc. The vernier scale has 12 divisions each side of zero. These are marked 0−60 minutes of arc, so that each division equals 1/12 of 60, that is 5 minutes of arc. Thus the reading of the vernier bevel protractor is:

(a) the largest 'whole' degree on the main scale indicated by the vernier zero division,

(b) the reading on the vernier scale in line with a main scale division.

Thus the reading shown on the scales in Fig. 3.15(c) is:

17 'whole' degrees	17°	00′
Vernier 25 mark in		
line with main scale	00	25′
	17°	25′

Providing the vernier protractor has been manufactured to the specifications of BS 1685, it will have an accuracy of 5 minutes of arc. However, as with all measuring instruments, its actual performance will depend upon the skill of the user. The scales should have a satin chrome finish to prevent corrosion and to improve the ease with which the scales can be read.

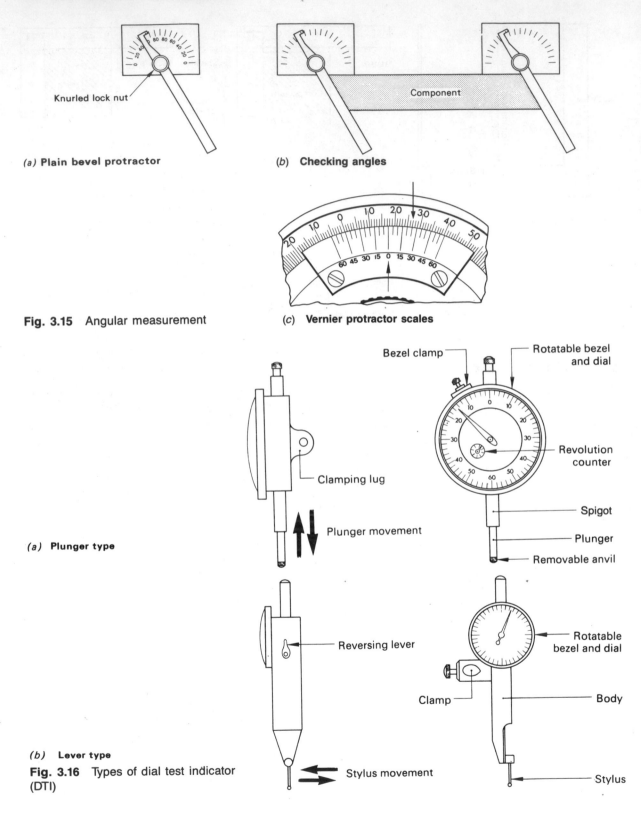

(a) **Plain bevel protractor**

(b) **Checking angles**

Fig. 3.15   Angular measurement

(c) **Vernier protractor scales**

(a) **Plunger type**

(b) **Lever type**

Fig. 3.16   Types of dial test indicator (DTI)

3.10 The dial test indicator

Essentially, the dial test indicator (DTI) measures the displacement of its plunger or stylus on a circular dial by means of a rotating pointer. Figure 3.16 shows the two types of this instrument normally used.

Plunger type

This type relies upon a rack and pinion to convert the linear movement of the plunger into rotary motion for the pointer. A gear train is used to magnify the movement of the plunger and, therefore, the rack and pinion. This type of instrument has a long plunger movement and is fitted with a secondary scale to count the number of revolutions of the main pointer. Various magnifications and dial marking are available.

Lever type

This type relies upon a lever and scroll system to magnify the movement of the stylus. This is shown in Fig. 3.16(*b*). It has only a limited range of stylus movement: little more than one revolution of the pointer. It is much more compact than the plunger type and is very popular for inspection and machine setting. An example of its use to eliminate errors of 'feel' when using the vernier height gauge is shown in Fig. 3.13.

Comparative measurement

The dial test indicator is not used for making direct measurements but is used to measure the *difference* between the size of a known standard and the corresponding dimension of a workpiece. This technique is called *comparative measurement* and the use of a dial test indicator as a *comparator* is shown in Fig. 3.17.

3.11 Principles of metal cutting

All metal cutting tools have to be ground to a basic wedge shape at the cutting edge. This shape is no accident, but is fundamental to the needs of metal cutting tools.

Clearance angle

One of the first controlled cutting operations we perform must be the sharpening of a pencil with a penknife. It is unlikely that any formal instruction is received before our first attempt. By trial and error we soon find that the knife blade must be presented to the wood at a definite angle if success is to be achieved, as shown in Fig. 3.18.

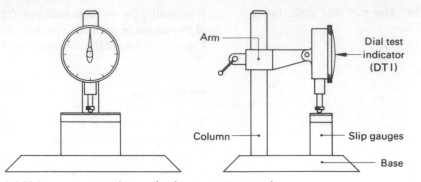

(a) Dial gauge mounted on a simple comparator stand

Arm

Dial test
indicator
(DTI)

Column

Slip gauges

Base

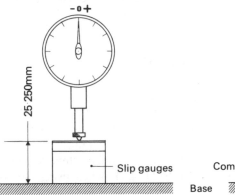

25 250mm

Slip gauges

Base

The dial gauge is set so that it reads zero
when slip gauges equal to the required
dimension are placed under the plunger

(b) Setting the comparator

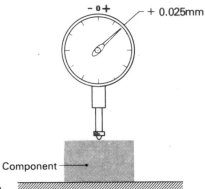

+ 0.025mm

Component

The component is placed under the
dial test indicator in place of the slip
gauges. Any error in the component
will be indicated on the D.T.I. as a ±
reading. In the example shown the
D.T.I. is reading 'plus' (+).
Therefore, the component is oversize

Fig. 3.17 Comparative measurement

(c) Making a comparative measurement

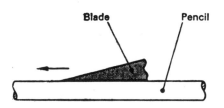

Blade   Pencil

(a) No clearance
*The blade skids along the
pencil without cutting*

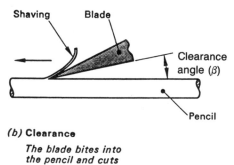

Shaving   Blade

Clearance
angle (β)

Pencil

(b) Clearance
*The blade bites into
the pencil and cuts*

Fig. 3.18 The clearance angle

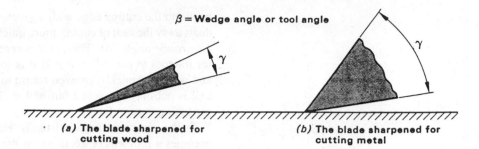

β = Wedge angle or tool angle

(a) **The blade sharpened for cutting wood**

(b) **The blade sharpened for cutting metal**

Fig. 3.19 Tool angle

Figure 3.18 immediately establishes the need for the angle β (beta) which is called the *clearance angle*. The clearance angle not only allows the tool to bite into the workpiece, but also reduces friction and, therefore, increases the life of the tool.

Tool angle

If, in place of a wooden pencil, a piece of soft metal such as brass is cut, it will be found that the cutting edge of the blade soon becomes blunt. If this edge is examined under a magnifying glass it will be seen that the cutting edge has crumbled away. For the blade to cut brass successfully, the cutting edge must be ground to a less acute angle to give greater strength, as in Fig. 3.19.

The angle to which the blade is ground is γ (gamma) and is called the *tool angle* or *wedge angle*.

The greater the tool angle, the stronger will be the tool. Unfortunately the greater angle γ is made, the greater will be the effort required to force the tool through the material being cut.

Rake angle

To complete the angles associated with cutting tools, reference must be made to angle α (alpha) which is called the *rake angle*. This is a very important angle, for it alone controls the geometry of the chip formation for any given material, and therefore controls the mechanics of the cutting action of the tool.

It is shown together with the angles previously considered in Fig. 3.20.

Summary

1. *Rake angle (α).* Increasing this angle makes cutting easier when ductile and low-strength materials are being cut, but reduces the strength of the cutting edge.
2. *Tool angle (γ).* Increasing the angle makes the tool stronger, but increases the required cutting force. Increasing the tool angle also

backs up the cutting edge with a greater mass of metal that conducts away the heat of cutting more quickly and prolongs tool life.

3. *Clearance angle (β)*. This angle is kept to the minimum required for the tool to cut (5° to 7°). If it is too small the tool will rub and wear out quickly, or even refuse to cut. If it is too large the tool is not only weakened but will tend to 'dig in' and 'chatter', producing a poor finish.

4. With the clearance angle virtually kept constant, tool design becomes a problem of deciding on the correct compromise between *rake angle* and *tool angle* that will give easy cutting combined with a satisfactory tool life.

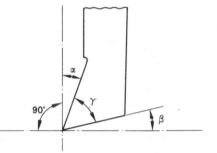

α (alpha) = Rake angle
γ (gamma) = Wedge or tool angle
β (beta) = Clearance angle

Fig. 3.20 Cutting tool angles

Some typical rake angles for high speed steel tools	
Material being cut	Rake
Cast iron	0°
Free-cutting brass	0°
Ductile brass	14°
Tin bronze	8°
Aluminium alloy	30°
Mild steel	25°
Medium carbon steel	20°
High carbon steel	12°
'Tufnol' plastic	0°

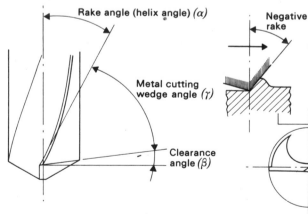

Rake angle (helix angle) *(α)*

Metal cutting wedge angle *(γ)*

Clearance angle *(β)*

(a) **Cutting angles applied to a twist drill**

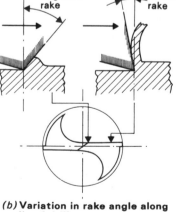

Negative rake

Positive rake

(b) **Variation in rake angle along lip of drill**

(*NOTE:* Rake angle at periphery is equal to helix angle of flute)

Fig. 3.21 The twist drill

Twist drill

This is dealt with in greater detail in Section 3.15. However Fig. 3.21(*a*) shows how the basic metal cutting wedge is applied to this cutting tool. Because the rake angle is formed by a helical groove, the rake angle varies from point to point along the lip of the drill, as shown in Fig. 3.21(*b*), from positive rake at the outer corner to negative rake at the bottom of the flute nearer the centre of rotation. The fact that cutting conditions are poor at the point of the drill does not affect the quality of the hole produced by the outer corners where cutting conditions are relatively good. The rake angle cannot be changed as it is controlled by the helix angle of the flute at the time of manufacture. However, the point angle and the clearance angle can be changed by grinding to suit the material being cut.

Lathe tools

Lathe tools come in a variety of shapes depending upon their application as shown in Fig. 3.22. Where the tool is cutting orthogonally, as in the case of the parting-off tool, the basic metal cutting wedge together with the rake and clearance angles are clearly defined as shown in Fig. 3.23(*a*). Typical values for these angles are the same as those given in Figure 3.20. Where the tool is cutting obliquely, as in the case of the straight nose roughing tool, the application of the cutting angles is rather more complex as they become compounded. The terminology of these angles is given in Fig. 3.23(*b*).

3.12 Power and hand tools

A strong man can exert approximately one tenth (1/10) of a kilowatt of power continuously. Compared with a power driven machine this is not very great and, in any case, our hero eventually tires whereas an electric motor continues to give its full output as long as the supply is connected. Compare the human output with the 5 kilowatt motor of a medium size centre lathe or the 15 to 50 kilowatts of a modern multi-spindle automatic production lathe. To use human energy for

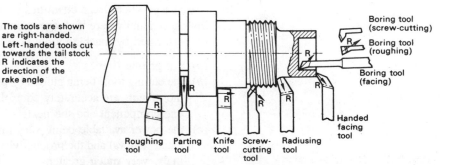

Fig. 3.22 Lathe tool profiles

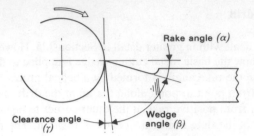

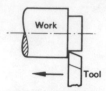

Rake angle (α)

Clearance angle (γ)

Wedge angle (β)

(a) **Cutting angles applied to an orthogonal turning tool**

Orthogonal cutting
The cutting edge is perpendicular to the direction of feed. Useful for producing a square shoulder at the end of a roughing cut

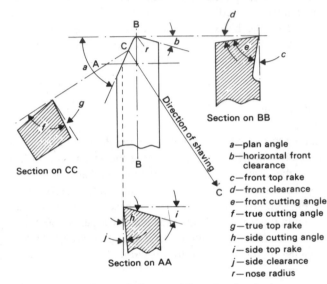

Section on BB

Section on CC

Section on AA

Direction of shaving

a—plan angle
b—horizontal front clearance
c—front top rake
d—front clearance
e—front cutting angle
f—true cutting angle
g—true top rake
h—side cutting angle
i—side top rake
j—side clearance
r—nose radius

Oblique cutting
The cutting edge is inclined to the direction of feed. Most efficient form for rapid metal removal

(b) **Cutting angles applied to an oblique turning tool**

Fig. 3.23 The lathe tool

extensive material removal is slow, expensive (compare wage rates per hour with the cost of electricity), and inefficient. Thus hand tools should only be used for finishing and fitting purposes, assembly and for small and delicate work which might be awkward to clamp securely in a machine or which might be damaged by such clamping. Hand tools are also used for prototype work where the cost of special tooling and work-holding devices is not warranted.

However where large quantities of material have to be removed at high speed and where high accuracy is required then the work must be performed by machine tools. These have the advantages of:

(a) not getting tired,
(b) the cutting tools being guided by precision slideways so that their movements are accurately controlled with no variation between one component and the next,
(c) the power available being very much greater so that the rate of metal removal and the productivity of the process is, correspondingly, very much greater.

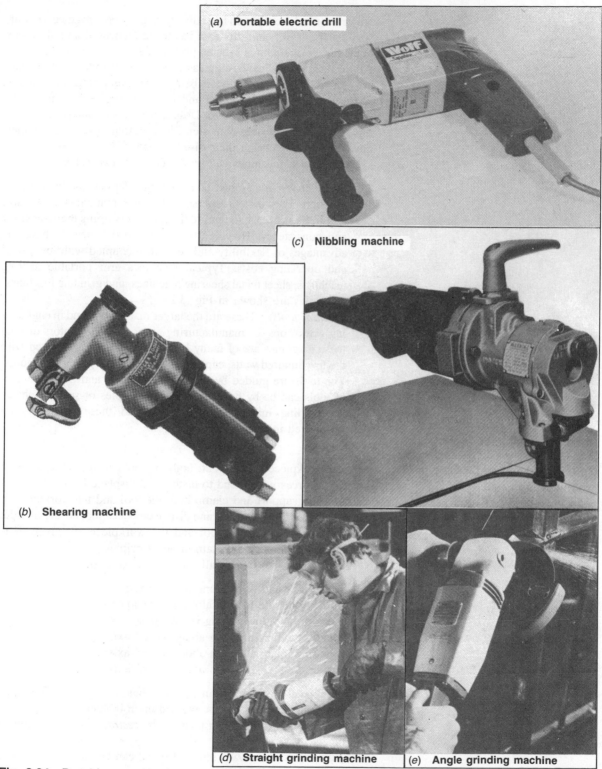

Fig. 3.24   Portable power tools

However, machine tools do have disadvantages. For instance, they are very costly to buy and this cost has to be recovered and allowance made for replacement at inflated prices when the machine wears out. All this is an additional cost on each component produced. Machines are not so 'flexible' as human beings and every time the work is changed expensive new tooling and work holding devices have to be made and fitted to the machines. Despite these disadvantages machines are still preferred where consistent, high output is required as their advantages far outweigh their disadvantages.

Power driven machines can be divided into two categories.

1. *Power driven hand tools*.   These reduce fatigue and increase the productivity of the craftsperson. However, manual skills are still required to control and guide them and this limits their accuracy and consistency. However, for small work, they do have the advantages of flexibility and versatility coupled with low initial and operating costs. Typical machines are: portable drilling machines, sheet metal shearing machines, and grinding machines. Examples are shown in Fig. 3.24.

2. *Machine tools*.   These are the larger machines found in engineering workshops for manufacturing purposes. The motors driving these machines are of many kilowatts' capacity, compared with the few hundred watts' capacity — or less — of power hand tools. The tools are guided by precision slideways and the movement of tools can be by hand or power. Examples of such machines are drilling machines and lathes, and these machines are introduced in Sections 3.14 and 3.15.

3.13 Restraints and locations when cutting

When a workpiece is being cut, both the tool and the workpiece are subject to forces which tend to distort and displace them. Thus it is necessary to support and clamp both the tool and the workpiece in such a manner that distortion and displacement cannot occur. In order to locate and restrain the tool and the workpiece effectively it is necessary to understand certain basic principles.

A body in space, free of all restraints is able to:

(*a*) move back and forth along its 'Y' axis;
(*b*) move from side to side along its 'X' axis;
(*c*) move up and down along its 'Z' axis;
(*d*) rotate in either direction about its 'Y' axis;
(*e*) rotate in either direction about its 'X' axis;
(*f*) rotate in either direction about its 'Z' axis.

Thus the metal block shown in Fig. 3.25 has *six degrees of freedom*.

In order that a body may be worked upon by hand or by machine it must be *located* in a given position by *restraining* its freedom of movement.

Figure 3.26 shows how a block of metal can be *located* in a given position by the application of suitable *restraints*. The base plate sup-

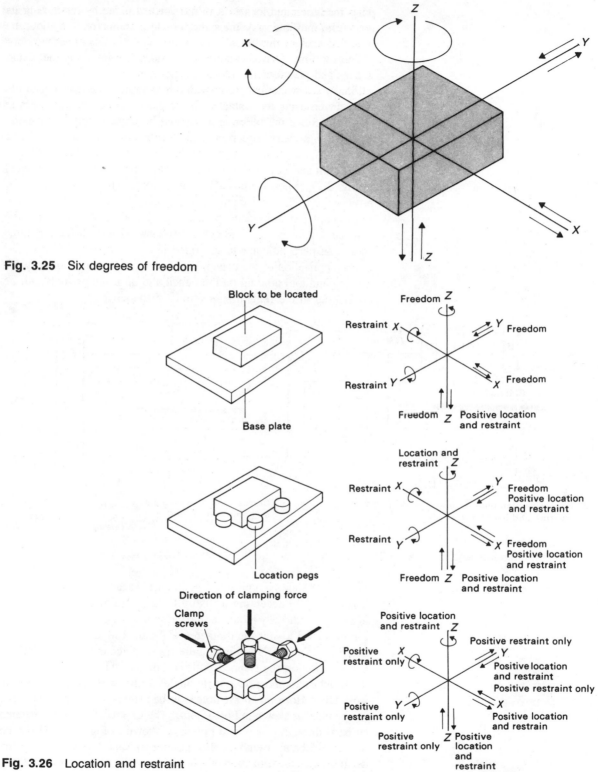

Fig. 3.25 Six degrees of freedom

Fig. 3.26 Location and restraint

ports the block and locates it in the vertical plane by restraining its downward movement. At the same time it restrains rotation about the X and Z axes of the block.

The addition of three location pegs adds restraint along the X and Z axes and positions the block on the plate.

Finally screw clamps are provided to complete the restraint of the block by ensuring its contact with the plate and the location pegs at all times. Since the block is restrained by contact with solid metal abutments in every direction it is said to be subjected to *positive* restraint.

Restraint may be *positive* or *frictional* and the difference is explained in Fig. 3.27. Wherever possible cutting forces should be resisted by positive restraints (solid abutments) and not by frictional restraint alone. For example, a component should be positioned in a vice so that the main cutting force is resisted by the fixed jaw. That is, the cutting force should be perpendicular to the fixed jaw and not parallel to it.

The application of the principles of restraint and location to work holding and tool holding on the bench and on machine tools will be considered in the remaining sections of this chapter.

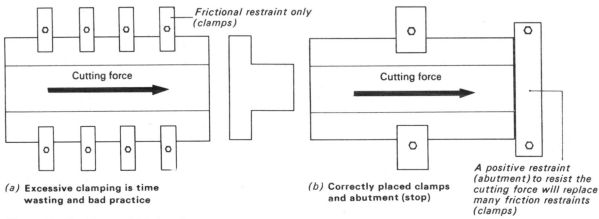

Frictional restraint only (clamps)

Cutting force

Cutting force

(a) **Excessive clamping is time wasting and bad practice**

(b) **Correctly placed clamps and abutment (stop)**

A positive restraint (abutment) to resist the cutting force will replace many friction restraints (clamps)

Fig. 3.27 Positive and frictional restraints (abutments and clamps)

3.14 The sensitive drilling machine

Figure 3.28 shows a typical bench mounted sensitive drilling machine. It is capable of accepting drills up to 12.5 millimetre diameter held in a chuck or directly mounted into the spindle nose. The spindle speed can be varied to suit the diameter of the drill used and the material being cut. This variation in spindle speed is achieved by altering the belt position on the stepped pulleys. For normal drilling operations the spindle axis must be at right angles to the working surface of the work table. However, if the hole is to be drilled at an angle, the table can be tilted as shown in Fig. 3.28(*b*). The feed mechanism is operated by hand through a rack and pinion as shown in Fig. 3.29. This type of feed mechanism enables the operator to 'feel' the progress of the drill as it cuts through the workpiece. Hence the name 'sensitive feed'.

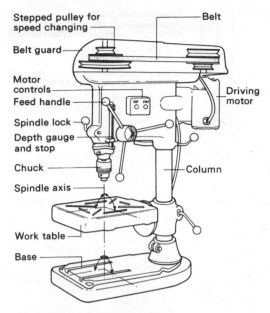

(a) **Bench drilling machine**

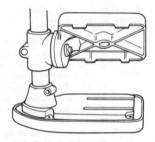

(b) **Table tilted**

Fig. 3.28 Sensitive bench drilling machine

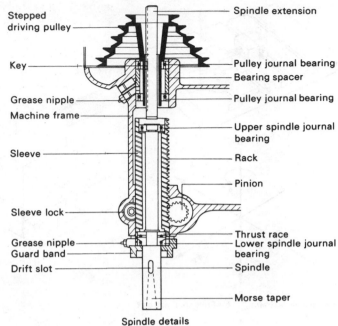

Spindle details

Fig. 3.29 Sensitive feed mechanism

To prevent accidents and damage to equipment the work being drilled must be securely clamped to the machine table, either directly or by means of a vice or other work holding device. Figure 3.30(*a*) shows the *restraints* acting upon a workpiece held in a machine vice whilst being drilled. The geometrical alignments necessary to *locate* the component correctly relative to the axis of the drilling machine spindle are shown in Fig. 3.30(*b*). Cylindrical components are more difficult to hold since only line contact exists between flat and round surfaces. It is advisable to insert a vee-block between the workpiece and the fixed jaw of the vice as shown in Fig. 3.31. Alternatively two vee-blocks may be used to support cylindrical work parallel to the machine table as shown in Fig. 3.32.

3.15   Twist drills and reamers

The twist drill does not produce a precision hole. Its sole purpose is to remove the maximum amount of metal as quickly as possible. The hole drilled is never smaller than the diameter of the drill, but is often larger as a result of incorrect point grinding. As well as dimensional inaccuracy, the hole is often out of round especially when opening up an existing hole using a two flute drill. Thus the drill should only be considered as a roughing out tool and, if a hole of accurate size, roundness, and good finish is required, then the hole should be drilled undersize and finish machined by means of a reamer or by single-point boring.

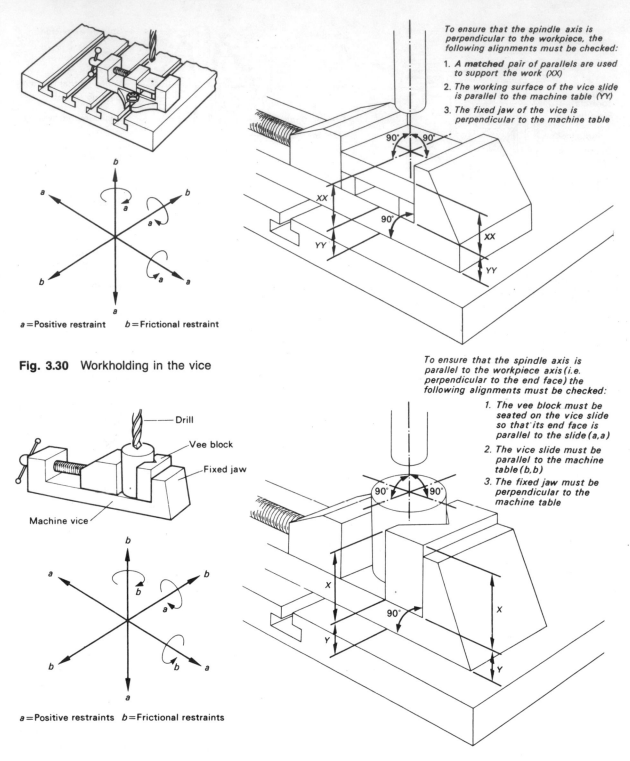

a = Positive restraint   *b* = Frictional restraint

Fig. 3.30 Workholding in the vice

To ensure that the spindle axis is perpendicular to the workpiece, the following alignments must be checked:

1. *A matched pair of parallels are used to support the work (XX)*

2. *The working surface of the vice slide is parallel to the machine table (YY)*

3. *The fixed jaw of the vice is perpendicular to the machine table*

To ensure that the spindle axis is parallel to the workpiece axis (i.e. perpendicular to the end face) the following alignments must be checked:

1. *The vee block must be seated on the vice slide so that its end face is parallel to the slide (a,a)*

2. *The vice slide must be parallel to the machine table (b,b)*

3. *The fixed jaw must be perpendicular to the machine table*

Drill
Vee block
Fixed jaw
Machine vice

a = Positive restraints   *b* = Frictional restraints

Fig. 3.31 Workholding cylindrical work in the vice

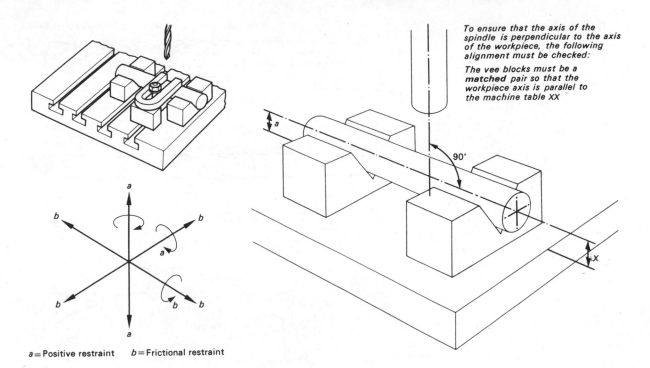

To ensure that the axis of the
spindle is perpendicular to the axis
of the workpiece, the following
alignment must be checked:

The vee blocks must be a
matched pair so that the
workpiece axis is parallel to
the machine table XX

a = Positive restraint b = Frictional restraint

Fig. 3.32 Workholding cylindrical
work on the drilling machine table

The modern twist drill is made from a cylindrical blank by machining two helical grooves in it to form the 'flutes'. These flutes run the full length of the drill body and have several functions:

(a) they provide the rake angle;
(b) they form the cutting edges;
(c) they provide a passage for the coolant;
(d) they facilitate swarf removal.

The flutes are not parallel to the axis of the drill but are slightly tapered. This produces a web which is thicker at the shank to give strength and thinner at the point to give better penetration of the 'chisel edge' when drilling from the solid. The lands are also ground with a slight taper so that the overall diameter of the drill at the shank is slightly less than its diameter at the point. This prevents the drill binding in the hole with a risk of breakage. Figure 3.33 shows a typical twist drill and names its more important features. The drill shown has a *taper shank* to fit directly into the spindle nose of the drilling machine. The same features and their names apply equally to parallel or (plain) shank drills which are held in a drill chuck attached to the drilling machine spindle nose.

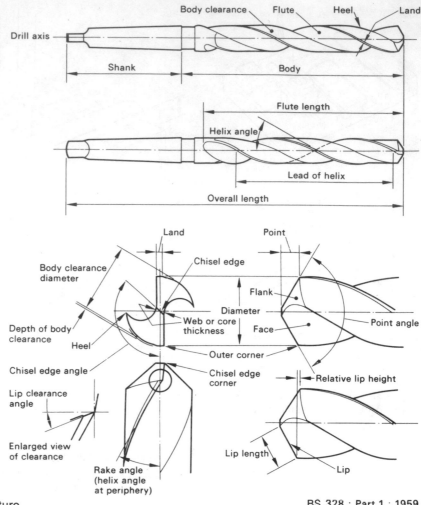

Fig. 3.33 Twist drill nomenclature

BS 328 : Part 1 : 1959

Tables 3.2 and 3.3 give typical cutting speeds and feed rates for use with standard high-speed steel twist drills under reasonably controlled conditions. If the speed or feed recommended is not available on the drilling machine, always select a speed or feed which is *less* than the recommended rate. Typical speed and feed calculations are given in examples 3.1 and 3.2.

Most drilling operations can be performed with a standard drill, however special drills are available with alternative point and helix angles. Examples are shown in Fig. 3.34 together with typical applications.

Twist drills suffer early failure or produce holes which are dimensionally inaccurate, out of round and of poor finish for the following general reasons:

(*a*) incorrect regrinding of the point;
(*b*) selection of incorrect feed and speeds;
(*c*) abuse and mishandling.

Table 3.2 Cutting speeds for HSS twist drills

Material being drilled	Cutting speed (m/min)
Aluminium	70–100
Brass	35–50
Bronze (phosphor)	20–35
Cast iron (grey)	25–40
Copper	35–45
Steel (mild)	30–40
Steel (medium carbon)	20–30
Steel (alloy-high tensile)	5–8
Thermo-setting plastic (Low speed due to abrasive properties)	20–30

Table 3.3 Feeds for HSS twist drills

Drill diameter (mm)	Rate of feed (mm/rev)
1.0–2.5	0.040–0.060
2.6–4.5	0.050–0.100
4.6–6.0	0.075–0.150
6.1–9.0	0.100–0.200
9.1–12.0	0.150–0.250
12.1–15.0	0.200–0.300
15.1–18.0	0.230–0.330
18.1–21.0	0.260–0.360
21.1–25.0	0.280–0.380

Example 3.1

Calculate the spindle speed in rev/min for a high-speed steel drill 12 mm diameter, cutting mild steel.

$$N = \frac{1000S}{\pi d}$$

where: N = spindle speed in rev/min
S = cutting speed in m/min
d = drill diameter (mm)
π = 3.14

From Table 3.2, a suitable cutting speed (S) for mild steel is 30 m/min, thus:

$$N = \frac{1000 \times 30}{3.14 \times 12}$$

796.2 rev/min (*Ans.*)

A spindle speed between 750 and 800 rev/min would be satisfactory.

Example 3.2

Calculate the time taken in seconds for the drill in Example 3.1 to penetrate a 15 mm thick steel plate.

From Example 3.1 the spindle speed has already been calculated as 796 rev/min (nearest whole number); from Table 3.3 it will be seen that a suitable feed for a 12 mm drill is 0.25 mm/rev.

$$t = \frac{60P}{NF}$$

where: t = time in seconds
P = depth of penetration (mm)
N = spindle speed (rev/min)
F = feed (mm/rev)

$$= \frac{60 \times 15}{796 \times 0.25}$$

= 4.5 seconds (*Ans.*)
(to one decimal place)

Table 3.4 summarises the more common causes of twist drill failures and faults and suggests probable causes and remedies. Because of their flexibility, twist drills are prone to 'wander' if the forces acting on the cutting edges of the point are out of balance. This is caused by the point not being ground symmetrically or the material being drilled not offering uniform resistance. Figure 3.35(*a*) shows how the drill is controlled by the chisel point when drilling from the solid. This results in the hole being oversize when the drill point is not ground symmetrically. However, when opening up existing holes the point is floating and the drill is guided only by its outer corners. This results in the hole being of the correct 'diameter' but out of round as shown in Fig. 3.35(*b*). This fault is overcome by using a three-flute 'core' drill. This type of drill is used for opening up cored holes in casting as well as opening up previously drilled holes. It cannot start a hole from the solid. Reamers will also correct the roundness of the hole.

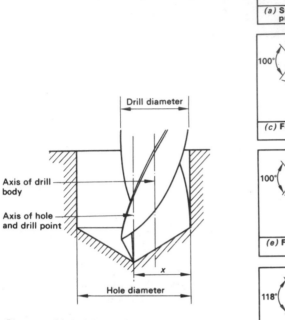

Diameter of hole drilled = 2x

where x = greatest distance

from drill point axis to a corner

Oversize hole caused by drill being ground off centre

(a) Oversize hole

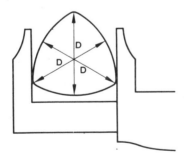

The 'lobed' constant diameter figure shown will give a constant reading when measured between two parallel surfaces

A hole of this shape will be produced when opening up an existing hole with a two flute drill

(b) The lobed figure

Fig. 3.35 Hole faults

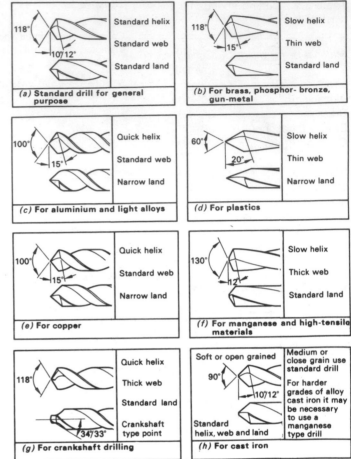

Fig. 3.34 Point angles

Figure 3.36 shows three types of machine reamer which may be used in a drilling machine. It can be seen that they have many more flutes and cutting edges than a standard drill or a core drill. The reamer is a *finishing tool* removing only very small amounts of metal but leaving a hole which is correct to size, round, and of good finish. Reamers are referred to as being *fluted* or *rose-action*. Note that machine reamers only have a bevel lead, whereas hand reamers also have a taper lead to guide them into the hole.

Fluted reamers cut on the end (bevel lead) only. The radial land (no clearance) prevents cutting taking place along the flutes. Fluted reamers give the best results with steels and similar ductile materials.

Rose-action reamers cut both on the end and on the periphery of the flutes, which are backed-off to give them a clearance angle. Such reamers are freer cutting but are not so accurate as the fluted reamers. Rose-action reamers are better for cast-iron, bronze, and plastic materials which tend to close on the reamer behind the cutting edge. The peripheral cutting action prevents seizure of the reamer in the hole under these conditions. Although standard reamers are made for right-hand cutting, they have flutes with a left-hand helix. This serves

Table 3.4 Twist drill fault-finding chart

Failure	Probable cause	Remedy
Damaged point	1. Do not use a hard-faced hammer when inserting the drill in the spindle 2. When removing the drill from the spindle, do not let it drop on to the hard surface of the machine table	Do not abuse the drill point
Rough hole	1. Drill point is incorrectly ground or blunt 2. Feed is too rapid 3. Coolant incorrect or insufficient	Regrind point correctly Reduce rate of feed Check coolant
Oversize hole	1. Lips of drill are of unequal length (Fig. 3.35) 2. Point angle is unequally disposed about drill axis 3. Point thinning is not central 4. Machine spindle is worn and running out of true	Regrind point correctly Recondition the machine
Unequal chips	1. Lips of drill are of unequal length 2. Point angle is unequally disposed about drill axis	Regrind point correctly
Split web (core)	1. Lip clearance angle too small 2. Point thinned too much 3. Feed too great	Regrind point correctly Reduce rate of feed
Chipped lips	1. Lip clearance angle too large 2. Feed too great	Regrind point Reduce rate of feed
Damaged corners	1. Cutting speed too high, drill 'blues' at outer corners 2. Coolant insufficient or incorrect 3. Hard spot, scale, or inclusions in material being drilled	Reduce spindle speed Check coolant Inspect material
Broken tang	1. Drill not correctly fitted into spindle so that it slips 2. Drill jams in hole and slips	Ensure shank and spindle are clean and undamaged before inserting Reduce rate of feed
Broken drill	1. Drill is blunt 2. Lip clearance angle too small 3. Drill point incorrectly ground 4. Rate of feed too great 5. Work insecurely clamped 6. Drill jams in hole due to worn corners 7. Flutes choked with chips when drilling deep holes	Regrind point Reduce rate of feed Re-clamp more securely Regrind point Withdraw drill periodically and clean

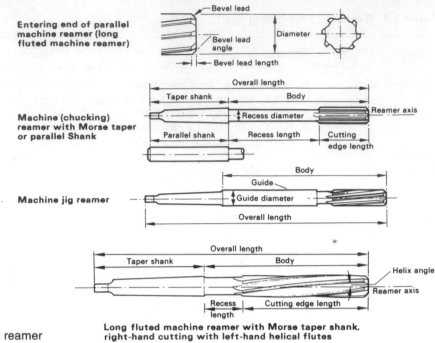

Fig. 3.36 Types of machine reamer

two purposes:

(*a*) to prevent the reamer being drawn into the hole by the 'screw action' of the helix;

(*b*) to eject the chips ahead of the reamer and prevent them being drawn back up the hole where they would spoil the finished surface.

The reamer always follows the axis of the existing hole: it cannot correct positional errors. If the original hole is out of position or out of alignment with its datum, then these errors must be corrected by single point boring.

3.16 Miscellaneous operations

In addition to drilling holes, the following operations are also performed on the drilling machine:

(*a*) trepanning;
(*b*) countersinking;
(*c*) counterboring;
(*d*) spot facing.

Trepanning

Not only is it dangerous to try and cut large diameter holes in sheet metal and thin plate with a twist drill, but the resultant hole will be unsatisfactory. There will be insufficient metal to guide the drill point and the drill will 'grab' and leave a hole which will be jagged, out of round, and oversize. In extreme cases 'grabbing' can tear the work clear from its restraints so that it rotates with the drill. This is not only extremely dangerous leading to severe injuries but can cause considerable and costly damage to the equipment.

One way to overcome this problem is to use a *trepanning cutter*. Instead of cutting away all the metal in the hole as swarf, the trepanning cutter merely removes a thin annulus of metal. This leaves a clean hole in the workpiece and a disc of metal slightly smaller than the hole as shown in Fig. 3.37(a). The simplest type of trepanning cutter is the traditional *tank cutter* shown in Fig. 3.37(b). This has a number of disadvantages, and the *hole saw* shown in Fig. 3.37(c) is superior where a number of holes of the same size have to be cut.

Countersinking

Figure 3.38(a) shows a typical *rose-bit* used for countersinking. Since the bit is conical in form it is self-centring and does not require a pilot to ensure axial alignment with the hole being countersunk. This process is used for the following purposes.

(a) To provide a recess for the head of a countersunk screw so that a flush surface is left after the screw is installed.
(b) To deburr a hole after drilling. Burred holes are dangerous to handle, and mating components with burred holes will not seat together correctly upon assembly.
(c) To chamfer sharp corners in order to make the component safer to handle and less likely to crack when hardened.

Counterboring

Figure 3.38(b) shows a typical *counterboring cutter*. It is similar in design to an end mill but is fitted with a pilot to ensure axial alignment with the hole being counterbored. Counterboring is used to provide a recess to accept the head of a cheese-head or a cap-head screw so that a safe, flush surface is left after installation of the screws. These screws are stronger than countersunk head screws.

Spot facing

Figure 3.38(c) shows a typical *spot facing cutter*. It is used to provide a flat seating on a raised boss for nuts or bolt heads to pull down onto otherwise rough cast or forged surfaces.

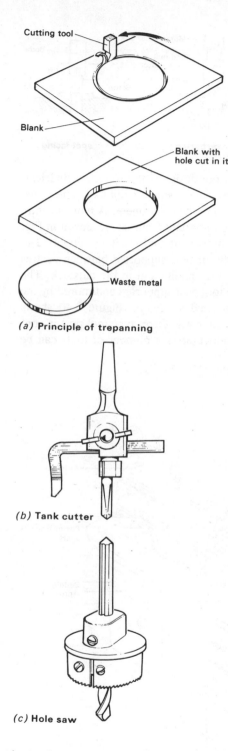

(a) Principle of trepanning

(b) Tank cutter

(c) Hole saw

Large diameter hole cutters for sheet-metal

Fig. 3.37 Trepanning large holes

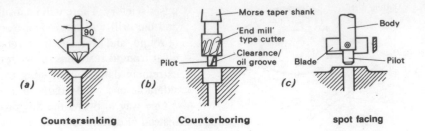

Fig. 3.38 Miscellaneous cutters

Countersinking   Counterboring   spot facing

3.17 The centre lathe

The centre lathe is a machine tool designed to produce cylindrical, conical and plane (flat) surfaces using a single point tool. These surfaces may be internal or external to the component. A typical centre lathe and the names of its more important features are shown in Fig. 3.39. The work is rotated by the spindle and is either supported in a chuck attached to that spindle or it is supported between centres which, in turn, are carried in the spindle and in the tailstock. The cutting tools are carried in the tool post supported and moved by the saddle (carriage) and its associated slideways. Figure 3.40 shows various types of tool post. The four-way turret type tool post is convenient for small quantity production as a number of tools can be

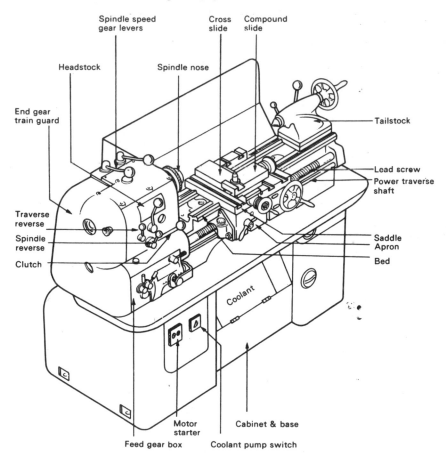

Fig. 3.39 The centre lathe

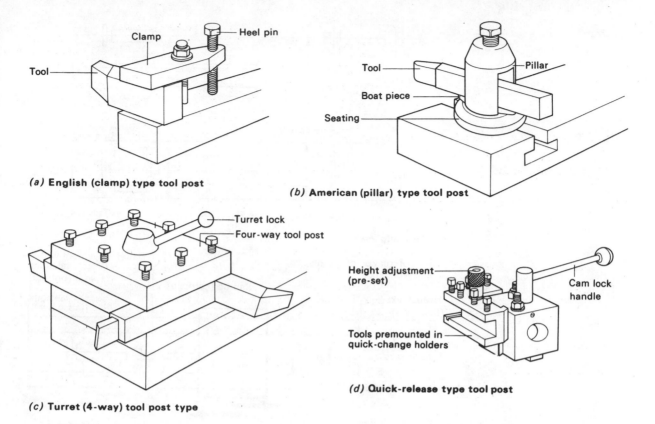

(a) **English (clamp) type tool post**

(b) **American (pillar) type tool post**

(c) **Turret (4-way) tool post type**

(d) **Quick-release type tool post**

Fig. 3.40 Centre lathe tool posts

mounted in it and swung into position as required, thus saving setting time. The Quick Release type also provides for the rapid changing of preset tooling. It has the advantage that it is not limited to four tools, and it also allows for easy height adjustment of the individual tools.

Figure 3.41(*a*) shows the basic alignment of the headstock, tailstock, spindle and bed slideways. It can be seen that the common spindle and tailstock axis is parallel to the bed slideways in both the vertical and horizontal planes. This is the *basic alignment* of the centre lathe and all other alignments are referred to it. The *movement of the saddle* (carriage) alone is shown in Fig. 3.41(*b*). It can be seen that it moves and guides the cutting tool in a path parallel to the spindle axis and this is the movement which produces *cylindrical surfaces* both externally and internally.

The *cross-slide* is on top of the saddle and is aligned at 90° to the spindle axis as shown in Fig. 3.41(*c*). Since the slide moves the tool in a path at right-angles to the spindle axis it is used for producing *plane surfaces*. This operation is called 'facing' and is used for producing the flat ends of turned components. The cross-slide is also used for controlling the 'in-feed' of the cutting tool when turning cylindrical surfaces from the saddle. That is, it controls the depth of cut.

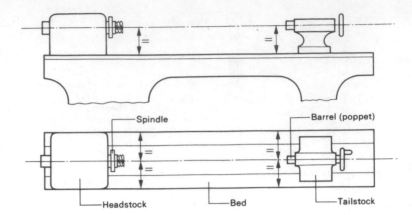

Spindle — — Barrel (poppet)

= =

= =

Headstock — Bed — Tailstock

(a) **Basic alignment**

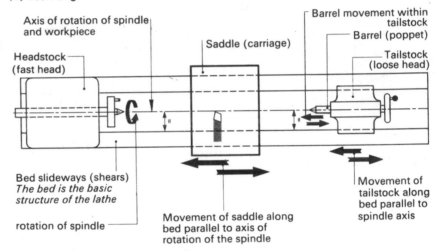

Axis of rotation of spindle and workpiece

Saddle (carriage)

Barrel movement within tailstock

Barrel (poppet)

Headstock (fast head)

Tailstock (loose head)

Bed slideways (shears)
The bed is the basic structure of the lathe

rotation of spindle

Movement of saddle along bed parallel to axis of rotation of the spindle

Movement of tailstock along bed parallel to spindle axis

(b) **The carriage or saddle provides the basic movement of the cutting tool parallel to the work axis**

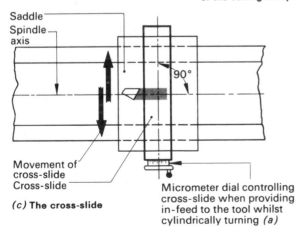

Saddle
Spindle axis

90°

Movement of cross-slide
Cross-slide

Micrometer dial controlling cross-slide when providing in-feed to the tool whilst cylindrically turning *(a)*

(c) **The cross-slide**

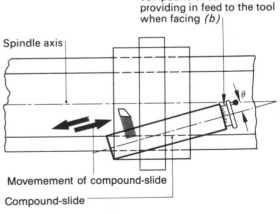

Micrometer dial controlling compound slide when providing in feed to the tool when facing *(b)*

Spindle axis

θ

Movemement of compound-slide

Compound-slide

(d) **The compound slide**

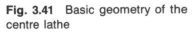

Fig. 3.41 Basic geometry of the centre lathe

The *compound-slide* or top-slide is located on top of the cross-slide and it can be set at an angle to the spindle axis as shown in Fig. 3.41(*d*). Since it moves the tool in a path which is at an angle to the spindle axis it is used to produce *conical* or tapered components. When set parallel to the spindle axis, the compound slide can be used to control the 'in-feed' or depth of cut of the tool when performing facing operations. Because both the cross- and compound-slides are used for controlling the depth of cut of the tool, they are both fitted with micrometer dials to their lead-screw hand wheels.

3.18   Work-holding on the centre lathe

The general requirements of lathework is that all diameters should be concentric or, if eccentricity is required, that the degree of offset should be accurately controlled. The most satisfactory way of achieving concentricity is to turn all the diameters at one setting. This is not always possible and a range of work-holding techniques and devices have been devised. These enable a wide range of components to be set and re-set to achieve the accuracy of concentricity desired. These work-holding devices must be capable of:

(*a*) locating the work relative to the spindle axis;
(*b*) rotating the work at the correct speed without slip;
(*c*) preventing the work being deflected by the cutting forces or its own weight: some slender work requires additional support (use of steadies);
(*d*) holding the work sufficiently rigidly so that it will not spin out of the machine, be ejected by the cutting forces, yet not be crushed or distorted by the work-holding device.

The following are the normal methods of work holding used on the centre lathe. They exclude special turning fixtures.

Between centres

This is the fundamental method of work-holding from which the centre lathe gets its name. The component is located between centres and driven by a catch plate and carrier as shown in Fig. 3.42(*a*). The restraints acting upon the workpiece are shown in Fig. 3.42(*b*), and

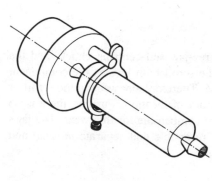

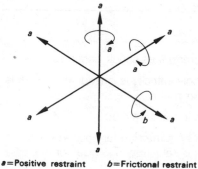

a=Positive restraint   *b*=Frictional restraint

(b) Restraints

Fig. 3.42 Workholding between centres

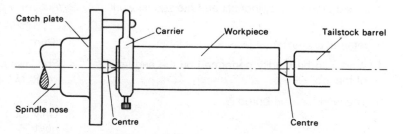

Catch plate — Carrier — Workpiece — Tailstock barrel — Spindle nose — Centre — Centre

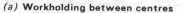
(a) **Workholding between centres**

Table 3.5 Work-holding between centres

Advantages	Limitations
1. Work can be easily reversed without loss of concentricity	1. Centre holes have to be drilled before work can be set up
2. Work can be taken from the machine for inspection and easily re-set without loss of concentricity	2. Only limited work can be performed on the end of the bar
3. Work can be transferred between machines (e.g. lathe and cylindrical grinder) without loss of concentricity	3. Boring operations cannot be performed
	4. There is lack of rigidity
4. Long work (full length of bed) can be accommodated	5. Cutting speeds are limited unless a revolving centre is used. This reduces accuracy and accessibility
	6. Skill in setting is required to obtain the correct fit between centres and work

the advantages and limitations of this method of workholding are listed in Table 3.5. When turning between centres, always take a trial cut along the component and test for parallelism. Any error can be corrected by lateral adjustment of the tailstock.

Self-centring chuck

Figure 3.43 shows a typical three-jaw self-centring chuck, and the restraints associated with it. The scroll both locates the component and provides the clamping force. Therefore any wear in the scroll or the jaws results in loss of accuracy of location. Further, there is no means of adjustment possible to compensate for this wear. The jaws for this type of chuck are *not* reversible and separate *internal* and

Table 3.6 The self-centring chuck

Advantages	Limitations
1. Ease of work setting	1. Accuracy decreases as chuck becomes worn
2. A wide range of cylindrical and hexagonal work can be held	2. Accuracy of concentricity is limited when work is reversed in the chuck
3. Internal and external jaws are available	3. 'Run out' cannot be corrected
4. Work can be readily performed on the end face of the job	4. Soft jaws can be turned up for second operation work, but this is seldom economical for one-off jobs
5. The work can be bored	5. Only round and hexagonal components can be held

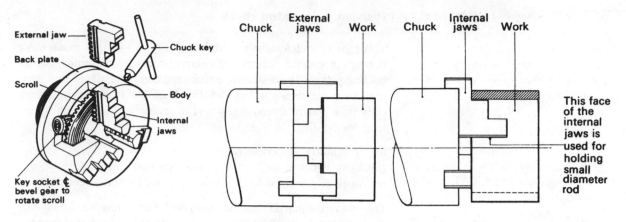

(a) **Construction**

(b) **Internal and external work-holding**

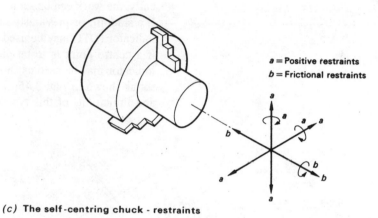

a = Positive restraints
b = Frictional restraints

Fig. 3.43 The three-jaw, self-centring chuck

(c) **The self-centring chuck - restraints**

external jaws have to be used as shown in Fig. 3.43(*b*). When changing jaws, the following precautions have to be observed as the jaws and the scroll are supplied as a matched set.

(*a*) Check that each jaw in the set carries the same serial number and that the serial number is the same as that on the chuck body.
(*b*) Insert the jaws sequentially, commencing with number 1 jaw in number 1 slot.

When new, self-centring chucks are quite accurate. To preserve this accuracy, *never*:

(*a*) try to hammer the work true;
(*b*) hold on a surface which is not cylindrical (e.g. hot-rolled (black) bar).
(*c*) Hold on the tips of the jaws.

The advantages and limitations of the three-jaw, self-centring chuck are listed in Table 3.6.

Four-jaw independent chuck

This type of chuck, which is shown in Fig. 3.44(*a*), is much more heavily constructed than the self-centring chuck and has much greater holding power. The restraints acting on a component held in a four-jaw chuck are shown in Fig. 3.44(*b*). Each jaw is moved independently by a square thread screw and is reversible. These chucks are used for holding:

(*a*) irregularly shaped work;
(*b*) work which must be trued up to run concentrically;
(*c*) work which must be deliberately offset to run eccentrically.

The jaws of the independent four-jaw chuck *can be reversed* for holding the work internally. Therefore separate internal and external jaws are not required with this type of chuck. Since the jaws move independently, the work can be set to run concentrically with the spindle axis. If a smooth, or previously machined surface, is available, a dial test indicator (DTI) may be used as shown in Fig. 3.45(*a*). Alternatively, if a centre point is to be picked up, a floating centre and dial test indicator may be used as shown in Fig. 3.45(*b*). Rough work may be set as shown in Fig. 3.45(*c*) using a scribing block. The advantages and limitations of this type of chuck are listed in Table 3.7.

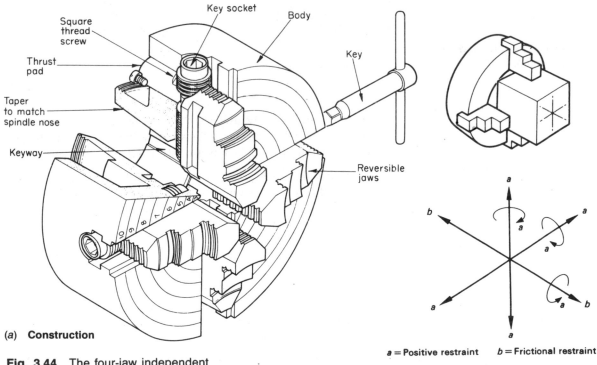

Key socket

Body

Square thread screw

Thrust pad

Taper to match spindle nose

Keyway

Key

Reversible jaws

(*a*) **Construction**

Fig. 3.44 The four-jaw independent chuck

a = Positive restraint   *b* = Frictional restraint

(*b*) **Restraints**

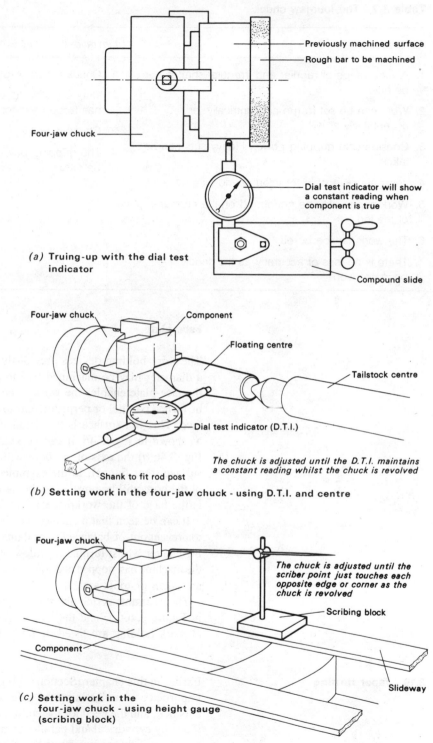

(a) **Truing-up with the dial test indicator**

Previously machined surface

Rough bar to be machined

Four-jaw chuck

Dial test indicator will show a constant reading when component is true

Compound slide

(b) **Setting work in the four-jaw chuck - using D.T.I. and centre**

Four-jaw chuck

Component

Floating centre

Tailstock centre

Dial test indicator (D.T.I.)

Shank to fit rod post

The chuck is adjusted until the D.T.I. maintains a constant reading whilst the chuck is revolved

(c) **Setting work in the four-jaw chuck - using height gauge (scribing block)**

Four-jaw chuck

The chuck is adjusted until the scriber point just touches each opposite edge or corner as the chuck is revolved

Scribing block

Component

Slideway

Fig. 3.45   The four-jaw chuck — work setting

Table 3.7 The four-jaw chuck

Advantages	Limitations
1. A wide range of regular and irregular shapes can be held	1. Chuck is heavy to handle on to the lathe
2. Work can be set to run concentrically, or eccentrically at will	2. Chuck is slow to set up. A dial test indicator (DTI) has to be used for accurate setting
3. Considerable gripping power. Heavy cuts can be taken	3. Chuck is bulky
4. Jaws are reversible for internal and external work	4. The gripping power is so great that fine work can be easily damaged during setting
5. Work can readily be performed on the end face of the job	
6. The work can be bored	
7. There is no loss of accuracy as the chuck becomes worn	

Face-plate

The work-holding devices previously described are designed so that a diameter may be machined true to an existing diameter. However, the face-plate enables the work to be located so that a diameter can be turned parallel or perpendicular to a previously machined flat surface. This flat surface is the datum from which the diameter is set as shown in Fig. 3.46. It can be seen that in the example shown in Fig. 3.46(*a*) the axis of the bore will be *perpendicular* to the datum surface (rear flange). In the example shown in Fig. 3.46(*b*) the axis of the bore will be *parallel* to the datum surface which, in this case, is the base of the workpiece.

It can be seen that a balance weight is used in the second example to prevent out-of-balance forces damaging the spindle bearings, causing chatter whilst turning, or allowing the work to swing round after the machine has stopped and trapping the operator whilst measurements are taken or tools are set.

The restraints acting upon work clamped to a face-plate are shown in Figure 3.46(*c*), and the advantages and restraints of this method of work holding are listed in Table 3.8.

3.19 Taper turning

Earlier in this chapter (Section 3.17) great stress was placed upon the importance of maintaining the axial alignment of the headstock and tailstock, and the need for the cutting tool to move parallel to this axis if a truly cylindrical and parallel component is to be produced. Consequently if these alignments are disturbed, so that the tool moves at an angle to the spindle-tailstock axis, a conical surface will be produced. *Taper turning* is the controlled disturbance of these alignments

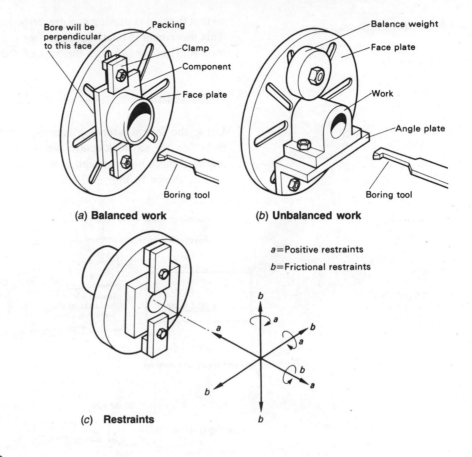

Fig. 3.46 The face-plate

(c) **Restraints**

Table 3.8 The face-plate

Advantages	Limitations
1. A wide range of regular and irregular components can be held	1. The face-plate is slow and tedious to set up. Not only must the workpiece be clocked up to run true, clamps must also be set up on the face-plate to retain the component
2. Work can be set to a datum surface. If the datum surface is parallel to the workpiece axis, it is set on an angle plate mounted on the face plate. If the datum surface is perpendicular to the workpiece axis, the workpiece is set directly on to the face plate	2. Considerable skill is required to clamp the component so that it is rigid enough to resist both the cutting forces, and those forces that will try to dislodge the work as it spins rapidly round
3. Work on the end face of the job is possible	3. Considerable skill is required to avoid distorting the workpiece by the clamps
4. The work can be bored	4. Irregular jobs have to be carefully balanced to prevent vibration, and the job rolling back on the operator
5. The work can be set to run concentrically or eccentrically at will	5. The clamps can limit the work that can be performed on the end face
6. There are no moving parts to lose their accuracy with wear	
7. The work can be rigidly clamped to resist heavy cuts	

so that the tool is moving at a prescribed angle to the workpiece axis. This movement is relative. It does not matter whether the path of the tool is offset, or whether the axis of the workpiece is offset.

Offset tailstock

Using the lateral adjusting screws, the body of the tailstock and, therefore, the tailstock centre can be offset. This inclines the axis of any workpiece held between centres relative to the path of the cutting tool as shown in Fig. 3.47(a).

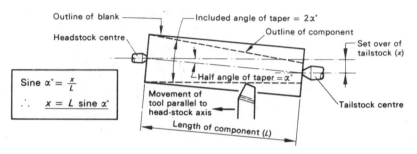

(a) **set over of centres**

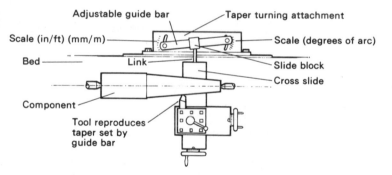

(b) **the taper turning attachment**

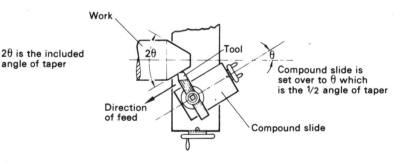

Fig. 3.47 Taper turning

(c) **compound slide**

Taper turning attachment

This is usually an 'optional-extra' which can be fitted to the rear of the lathe. It carries a guide bar which can be adjusted through a narrow range of angles. A slider on this bar is connected by a link to the cross-slide. As the saddle traverses along the bed of the lathe the link moves the cross-slide and, therefore, the tool in a path parallel to the guide bar of the taper turning attachment as shown in Fig. 3.47(*b*). The handwheel on the cross-slide can still be used for adjusting the in-feed of the tool to control the depth of cut.

Compound-slide

Setting over the compound-slide is the simplest method of producing a tapered surface. It is usually used for short, acute, tapers as shown in Fig. 3.47(*c*).

The advantages and limitations of all these techniques are listed in Table 3.9.

When turning any workpiece, but particularly when taper turning, it is very important to have the tool point in line with the centre height of the work. Not only are the cutting angles seriously affected by mounting the tool off-centre but also tapered work will not be truly conical. In fact if a tool is raised too far above centre the clearance angle becomes non-existent and the tool will not cut, whilst if the tool

Table 3.9 Comparison of taper turning techniques

Method	Advantages	Limitations
Set over of tailstock	1. Power traverse can be used 2. The full length of the bed can be used	1. Only small angles can be accommodated 2. Damage to the centre holes can occur 3. Difficulty in setting up 4. Only applies to work held between centres
Taper turning attachments	1. Power traverse can be used 2. Ease of setting 3. Can be applied to chucking and centre work	1. Only small angles can be accommodated 2. Only short lengths can be cut (304–457 mm (12–18 in) depending on make)
Compound slide	1. Very easy setting over a wide range of angles. (Usually used for short steep tapers and chamfers) 2. Can be applied to chucking and centre work	1. Only hand traverse available 2. Only very short lengths can be cut. Varies with m/c but is usually limited to about 76–101 mm (3–4 in)

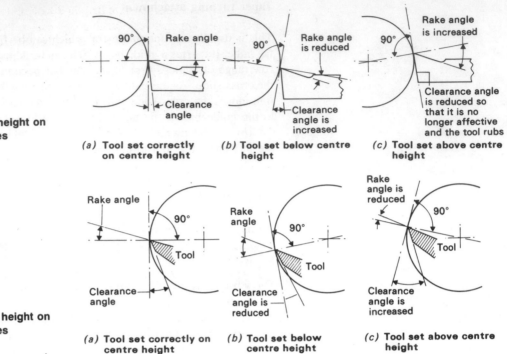

(a) **Effect of tool height on turning tool angles**

(a) **Tool set correctly on centre height**

(b) **Tool set below centre height**

(c) **Tool set above centre height**

(b) **Effect of tool height on boring tool angles**

(a) **Tool set correctly on centre height**

(b) **Tool set below centre height**

(c) **Tool set above centre height**

Fig. 3.48   Effect of tool height

is too low the work will tend to climb over the tool causing serious damage to the equipment and possible injury to the operator. Figure 3.48(*a*) shows the effects on the tool angles of setting tools above and below centre compared with a tool set correctly. Figure 3.48(*b*) shows the effects for the incorrect setting of boring tools.

3.20   Producing hollow components on the lathe

Hollow as well as solid components have to be produced on the lathe. Holes, concentric with the tool axis, may be produced by:

(*a*) drilling;
(*b*) reaming;
(*c*) single-point boring;
(*d*) a combination of the above.

The technique or combination of techniques chosen will depend upon the size of the hole and its depth/diameter ratio, the accuracy required, and whether a 'through' or a 'blind' hole is required.

Twist drill

Unless the bore is only a shallow recess, it is usual to rough out the hole with a twist drill held in the tailstock. A start is made for the twist drill by means of a centre drill in exactly the same way as when preparing a component for turning between centres. Since the feed force that can be exerted on the drill by the tailstock is limited, it is usual to gradually open up the hole to the required size using successively larger drills. The limitations of drilled holes are:

(*a*) poor finish compared with boring or reaming;
(*b*) lack of dimensional accuracy;
(*c*) lack of 'roundness' or geometrical accuracy;
(*d*) lack of positional accuracy as the drill tends to 'wander', especially when drilling deep holes in a soft, weak material such as brass.

Reamer

The quality of the hole is greatly improved if it is drilled slightly undersize and finished with a reamer (Section 3.15). The reamer should be held in a 'floating' holder so that it can follow the drilled hole without flexing. This prevents ovality and bell-mouthing. Figure 3.49 shows a typical reamer holder. A reamed hole has a good finish and a high degree of roundness. However the limitations of a reamed hole are:

(*a*) lack of positional accuracy since the reamer follows the axis of the original drilled hole and reproduces any 'wander' that is present;
(*b*) unless the quantity of components being produced warrants special tooling, only holes whose diameters are the same as standard reamer sizes can be produced.

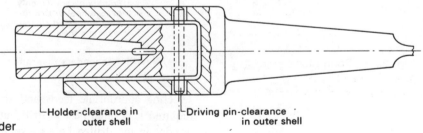

Holder-clearance in outer shell   Driving pin-clearance in outer shell

Fig. 3.49 Floating reamer holder

Boring

Figure 3.50 shows various types of boring tools and boring bars. It also shows the need for *secondary clearance* on a boring tool to prevent the 'heel' of the tool rubbing on the finished surface of the bore. Because of the overhang of the tool point from the tool-post and the slender shank of the tool, boring tools are prone to chatter and the cut is also liable to 'run-off' due to the deflection of the tool shank. This makes boring a highly skilled operation compared with turning external diameters, and great care is required in grinding the tool and

(a) **Solid bottoming tool for blind holes**

(b) **Solid roughing tool for through hole**

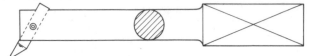

(c) **Boring bar with inserted tool bit for bottoming a blind hole**

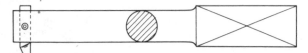

(d) **Boring bar with inserted tool bit for roughing a through hole**

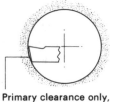

Primary clearance only, causes heel of tool to rub. Increasing clearance weakens the tool

secondary clearance prevents heel of tool rubbing, but permits a small primary clearance to be used to give strength to the tool

(e) **Need for secondary clearance**

Fig. 3.50 Boring tools

selecting appropriate feeds and speeds. If a standard size hole is required it is often best to drill out most of the metal; remove any wander in the drilled hole by single-point boring, leaving the hole slightly undersize; and finish the hole to size by reaming to ensure accurate size, roundness and a good finish.

Boring is the only possible means of removing 'wander' in the hole axis and giving a high degree of positional accuracy, but it suffers from the following limitations.

(a) Chatter and poor surface finish, especially on small diameter holes where only a slender tool can be used;

(b) The hole tends to be oval and bell-mouthed due to deflection of the boring tool shank. These defects become less severe as the hole diameter increases, thus allowing a boring tool or bar with a thicker and more rigid shank.

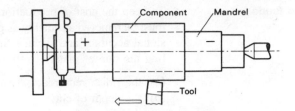

(a) Turning on the mandrel

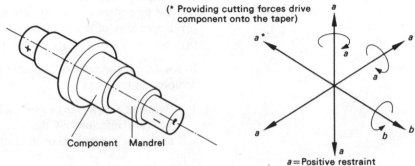

Fig. 3.51 Work-holding on the mandrel

(* Providing cutting forces drive component onto the taper)

a* Positive restraint
b = Frictional restraint

Where a hole is too small to be bored accurately, it is usual to drill and ream the component so as to produce a hole which is correct dimensionally and geometrically. This hole then becomes the datum, and the component is mounted on a mandrel as shown in Fig. 3.51. The external diameters can then be turned concentric with the bore. The mandrel is tapered so that the further the component is forced on, the more firmly it is held in place. Therefore the direction of cutting should always be towards the 'plus end' of the mandrel. The advantages and limitations of using a mandrel are listed in Table 3.10.

Table 3.10 Work-holding on the mandrel

Advantages	*Limitations*
1. Small-bore components can be turned with the bore and outside diameters concentric	1. Bore must be a standard size to fit a taper mandrel. Adjustable mandrels are available but these tend to lack rigidity and accuracy
2. Batch production is possible without loss of concentricity or lengthy set-up time	2. Cuts should only be taken towards the 'plus end' of the mandrel
3. The advantages of turning between centres apply (see Table 3.5)	3. Only friction drive available, and this limits size of cut that can be taken
	4. Special mandrels can be made but this is not economical for one-off jobs

3.21 Speeds and feeds

To keep the cost of production to a minimum, the surplus material must be removed as quickly as possible, providing the required dimensional accuracy and surface finish is achieved. The factors controlling the rate of material removal are:

(*a*) the finish required,
(*b*) the depth of cut,
(*c*) the rate of feed,
(*d*) the tool geometry,
(*e*) the properties and rigidity of the cutting tool and its mounting,
(*f*) the properties of the workpiece material,
(*g*) the rigidity of the workpiece,
(*h*) the power and rigidity of the machine tool.

In a machining operation the same rate of material removal may be achieved in two ways.

1. By using a high rate of feed and a shallow cut as shown in Fig. 3.52(*a*). Unfortunately this not only leads to a rough finish, but imposes a greater load on the cutting tool as shown in Fig. 3.52(*b*).

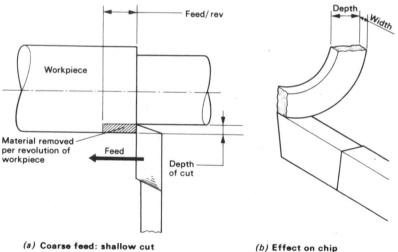

(a) **Coarse feed: shallow cut**

(b) **Effect on chip**

With coarse feed and shallow cut, the chip is bent across its deepest section. The bending force increases as the cube of the depth of the chip, i.e. doubling the depth of the chip increases the bending force eight times

Fig. 3.52 Effect of high feed rates

2. By using a low rate of feed and a deep cut as shown in Fig. 3.53(*a*). This gives better finish and, if the size is correct, avoids the need for taking a finishing cut. It also reduces the load on the cutting tool as shown in Fig. 3.53(*b*). Unfortunately a deep cut at a low feed rate can lead to chatter and a poor surface finish. Therefore a compromise has to be reached between depth of cut and rate of feed to suit any given set of conditions.

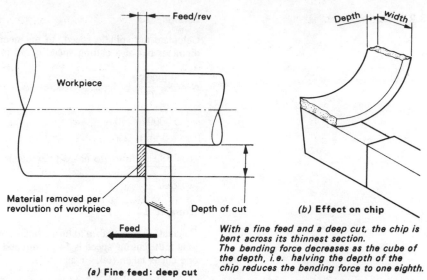

Fig. 3.53 Effect of deep cuts

(a) **Fine feed: deep cut**

(b) **Effect on chip**

With a fine feed and a deep cut, the chip is bent across its thinnest section.
The bending force decreases as the cube of the depth, i.e. halving the depth of the chip reduces the bending force to one eighth.

Table 3.11 Cutting speeds and feeds for HSS turning tools

Material being turned	Feed (mm/rev)	Cutting speed (m/min)
Aluminium	0.2–1.00	70–100
Brass (alpha) (ductile)	0.2–1.00	50–80
Brass (free-cutting)	0.2–1.5	70–100
Bronze (phosphor)	0.2–1.0	35–70
Cast iron (grey)	0.15–0.7	25–40
Copper	0.2–1.00	35–70
Steel (mild)	0.2–1.00	35–50
Steel (medium carbon)	0.15–0.7	30–35
Steel (alloy-high tensile)	0.08–0.3	5–10
Thermo-setting plastic (low speed due to abrasive properties)	0.2–1.0	35–50

Notes:
1. The above feeds and speeds are for ordinary HSS tools. For *super* HSS tools the feeds would remain the same, but the cutting speeds could be increased by 15% to 20%.
2. The *lower* speed range is suitable for heavy, roughing cuts.
   The *higher* speed range is suitable for light, finishing cuts.
3. The feed is selected to suit the finish required, and rate of metal removal.

Table 3.11 gives suitable cutting speeds and feeds for turning operations on a centre lathe using high-speed steel tools. These are only a guide, and the actual rates used may be increased or decreased, as experience dictates, for any particular set up. Typical calculations are given in examples 3.3 and 3.4.

Example 3.3

Calculate the spindle speed, to the nearest rev/min, for turning a 25 mm diameter bar at a cutting speed of 30 m/min (take π as 3.14).

$$N = \frac{1000S}{\pi D}$$

where: N = spindle speed
S = 30 m/min
π = 3.14
D = 25 mm

$$= \frac{1000 \times 30}{3.14 \times 25}$$

$$= \underline{382 \text{ rev/min}} \text{ (to nearest rev/min)}$$

Example 3.4

Calculate the time taken to turn a brass component 49 mm diameter by 70 mm long if the cutting speed is 44 m/min and the feed is 0.5 mm/revolution. Only one cut is taken (take π as 22/7).

$$N = \frac{1000S}{\pi D}$$

where: N = spindle speed
S = 44 m/min

$$= \frac{1000 \times 44 \times 7}{22 \times 49}$$

$$\pi = \frac{22}{7}$$

$$= 286 \text{ rev/min}$$
(to nearest rev/min)

$$D = 49 \text{ mm}$$

Rate of feed
$$= 0.5 \text{ mm/rev}$$
$$= 0.5 \times 286 \text{ mm/min}$$
$$= 143 \text{ mm/min}$$

Time taken to traverse 70 mm
$$= \frac{70}{143} \text{ min}$$

$$= \frac{70 \times 60}{143}$$

$$= \underline{29.37 \text{ s}}$$

3.22 Cutting fluids

Cutting fluids are designed to fulfil one or more of the following functions:

(*a*) to cool the tool and workpiece;
(*b*) to lubricate the chip/tool interface and reduce tool wear due to friction;
(*c*) to maintain a barrier between the chip and the tool face and thus prevent chip welding taking place and causing a built-up edge;
(*d*) to improve the surface finish of the workpiece;
(*e*) to flush away the chips (swarf);
(*f*) to prevent corrosion of the work and the machine.

There are many types and applications of cutting fluid and it is only possible to consider the general principles in the scope of this chapter. It is always best to consult the expert advisory service offered by the coolant manufacturers, as the correct selection and use of cutting fluids can increase the productivity of a machine shop far more cheaply than any other method, the costs involved being more than offset by the increased output and lower tool maintenance and replacement costs.

Mineral lubricating oils are unsuitable as cutting fluids as they give off clouds of unpleasant smoke and fumes when they come into contact with the hot swarf. They also have poor cooling properties and their lubricating properties are also unsuitable. The chip and tool have a small contact area, therefore the load per unit area on the lubricant is much greater than in a bearing. The *film strength* of a mineral lubricating oil is insufficient to withstand this pressure and the chip 'punctures' the oil film and comes into contact with the tool face and lubrication ceases to exist. Paraffin (kerosene) used to be used for machining aluminium and its alloys. However, because of the fire hazard and fume problems, it has been largely superseded by special water-based emulsified coolants.

Emulsified oil (suds)

When water and oil are added together they refuse to mix, but if an emulsifier (detergent) is added the oil will break up into droplets which spread throughout the water. This is what happens when the so called 'soluble' oils are added to water. The milky appearance of these *emulsions* is due to light being refracted by the oil droplets. The workshop name of 'suds' is due to this milky appearance. Since the oil is highly diluted with water, these emulsions are very cheap to use and form the most widely-used group of cutting fluids to be found in the machine shop.

The dilution with water reduces the lubricating properties and, as a result, soluble oils are not suitable for the very severe conditions found on many automatic machines or on machines using form cutters such as broaching and gear cutting machines. However, the high water content makes emulsified oils excellent coolants and for the general machine shop they are ideal, especially for manually operated machines taking fast but light cuts, and where the operator would be affected by the fumes given off by 'straight' (undiluted) cutting oils. The dilution with water helps to reduce the incidence of allergic skin complaints. The inclusion of a disinfectant helps to prevent the transmission of disease from one operative to another, and also helps to prevent the emulsion breaking down under bacteriological attack when it is standing.

Problems (marking out)

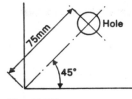

Fig. 3.54

Section A

1 A cellulose lacquer is painted on to a surface prior to marking out in order to:
   (a) prevent corrosion;
   (b) provide a means of identification;
   (c) provide a contrasting background;
   (d) protect the scriber point.

2 Hermaphrodite (odd-leg) calipers are used to scribe
   (a) circular lines;
   (b) lines parallel to an edge;
   (c) lines perpendicular to an edge;
   (d) irregular profiles.

3 Scribing instruments should be kept sharp with the aid of:
   (a) a grinding wheel;
   (b) a smooth file;
   (c) metal polish;
   (d) an oil stone.

4 The hole shown in Fig. 3.54 has been dimensioned using:
   (a) polar coordinates;
   (b) rectangular coordinates;
   (c) a centre line datum;
   (d) an edge datum.

5 A fixed point, line, edge or surface from which a measurement can be taken is known as a:
   (a) coordinate;
   (b) centre;
   (c) datum;
   (d) dimension line.

Section B

6 (a) Figure 3.55 shows a casting, the faces of which marked ▽ have been machined. Describe with the aid of sketches:

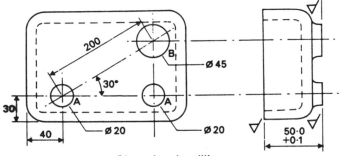

Dimensions in millimetres
Bosses A are of equal diameter
Material: grey cast iron

Fig. 3.55

(i) the method of locating and holding the casting relative to the datum surface;

(ii) the method of preparing the boss surfaces so that the scribed lines will show up clearly;

(iii) the process of marking out the boss centres.

(b) Calculate the rectangular coordinates for the bosses.

7 (a) Explain why it is usual to mark the position of a scribed line with a series of fine dots.

(b) Explain, with the aid of sketches, how dimensioning a row of holes from an edge datum results in a smaller cumulative error than 'chain' dimensioning.

8 Show, with the aid of sketches, how a universal surface gauge can be used for:

(a) scribing lines parallel to a datum surface;

(b) scribing lines parallel to an edge;

(c) setting surfaces parallel to a datum surface;

(d) carrying a dial test indicator to check the parallelism of a surface.

9 Describe *three* reasons for marking out a component before manufacture and describe how a sawn mild steel blank 100 mm × 75 mm × 12 mm should be prepared for marking out on the flat faces, using two of the edges as datums.

10 Describe how the following marking out processes are performed. Your answers should contain clear sketches.

(a) Finding the centre of a bar using the 'four arc' technique;

(b) finding the centre of a bar using the centre finder from a combination set;

(c) boxing a hole prior to drilling;

(d) constructing a perpendicular line using trammels;

(e) stepping off equidistant hole centres round a pitch circle.

Problems (measuring)

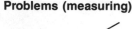
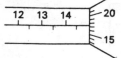

Fig. 3.56

Section A

1 The reading of the metric micrometer scales shown in Fig. 3.56 is:

(a) 14.77 mm;

(b) 14.67 mm;

(c) 14.72 mm;

(d) 15.17 mm.

2 The most accurate standard of length normally available in the workshop is the:

(a) rule;

(b) micrometer caliper;

(c) vernier caliper;

(d) slip (block) gauge.

3 To avoid reading errors due to parallax an engineer's rule should be

(a) as thin as possible;

(b) as thick as possible;

(c) satin chrome finished;

(d) engine engraved.

4 When using a micrometer a constant measuring pressure can be obtained by using the:
 (*a*) thimble;
 (*b*) spindle;
 (*c*) ratchet;
 (*d*) barrel.

5 The vernier scale shown in Fig. 3.57 has a reading accuracy of 1/40 mm. What is the reading shown?
 (*a*) 44.00 mm;
 (*b*) 44.15 mm;
 (*c*) 44.60 mm;
 (*d*) 49.15 mm;

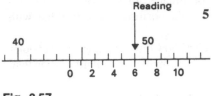

Reading

Fig. 3.57

Section B

6 (*a*) State four features that should be checked when selecting a good quality engineer's rule.
 (*b*) Describe with the aid of sketches how internal and external diameters can be measured using a rule and calipers.
 (*c*) State five precautions that should be observed when using calipers to ensure safe and accurate measurements.

7 (*a*) Draw the scales of a 25−50 mm micrometer caliper so that they show a reading of 43.74 mm.
 (*b*) Draw the scales of a 50 division vernier caliper so that they show a reading of 150.28 mm.
 (*c*) Draw the scales of a 0−25 mm depth micrometer so that they show a reading of 15.63 mm.

8 (*a*) Describe in detail the principle of operation of the micrometer.
 (*b*) Explain what precautions should be taken to maintain the high initial accuracy of a micrometer caliper.
 (*c*) Explain, with the aid of sketches, how a micrometer caliper should be adjusted to remove initial zero error.

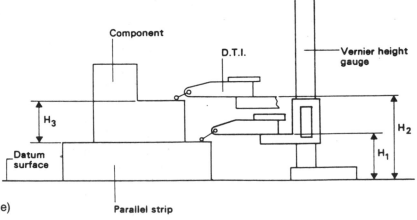

Fig. 3.58 Checking milled component (vernier height gauge)

9 (*a*) Sketch a typical vernier protractor and draw, in good proportion, an enlarged view of the scales to show a reading of 13° 15′.

(*b*) Sketch a good quality engineer's try-square and name the parts. Explain, with the aid of sketches, how a try-square may be checked for perpendicularity.

10 (*a*) With reference to Fig. 3.58, calculate the height of the step (H_3) if the reading of $H_1 = 50.82$ mm and the reading of $H_2 = 96.24$ mm.

(*b*) Explain the purpose of the dial test indicator in this application of the vernier height gauge and the need for a datum surface.

Problems (material removal)

Section A

1 The metal cutting wedge is fundamental to the geometry of:
   (*a*) hand tools only;
   (*b*) power-driven tools only;
   (*c*) sheet metal cutting tools only;
   (*d*) all cutting tools.

2 Continuous chips are formed when cutting:
   (*a*) ductile materials;
   (*b*) brittle materials;
   (*c*) amorphous plastic materials;
   (*d*) free-cutting non-ferrous alloys.

3 When clamping a workpiece, ready for machining, the main cutting force should be resisted by:
   (*a*) frictional restraint;
   (*b*) a clamp;
   (*c*) a solid abutment;
   (*d*) a spring-loaded abutment.

4 The rake angle of a cutting tool:
   (*a*) prevents rubbing;
   (*b*) controls the chip formation;
   (*c*) determines the profile of the tool;
   (*d*) determines whether the cutting action is oblique or orthogonal.

5 The 'suds' used as a coolant in general machine shops consists of:
   (*a*) a solution of detergent and water;
   (*b*) a straight mineral oil;
   (*c*) an emulsion of oil and water;
   (*d*) a chemical solution.

Section B

6 (*a*) Draw a simple single point metal cutting tool and indicate: (i) the rake angle; (ii) the clearance angle; (iii) the wedge angle.

(*b*) With the aid of a diagram show how varying the rake angle affects the length of the shear plane when cutting.

7 (a) Show, with the aid of sketches, what is meant by the terms *oblique* cutting and *orthogonal* cutting and the effect they have on the cutting action of the tool.

(b) Describe the two basic types of chip formed when cutting metals. Name two metals in each case which will form the type of chip discussed.

8 With the aid of clear sketches show how rake and clearance angles are applied to the following metal cutting tools;
(a) hacksaw;
(b) chisel;
(c) thread cutting tap;
(d) twist drill;
(e) lathe parting-off tool;
(f) scraper.

9 (a) Describe the main function of a cutting fluid and compare the advantages and limitations of: (i) straight fatty oils; (ii) compounded (blended) oils; (iii) emulsified oils (suds).

(b) Describe how emulsified oils should be mixed and stored to ensure stability and effective use.

10 Describe in detail the precautions that should be taken when resharpening cutting tools on an off-hand grinding machine.

Problems (drilling and drilling machines)

Section A

1 The spindle speed for a 10 mm diameter twist drill cutting at 30 m/min is:
(a) 942 rev/min;
(b) 955 rev/min;
(c) 1047 rev/min;
(d) 3000 rev/min.

2 The time taken to drill through a 10 mm thick plate at 300 rev/min and a feed of 0.2 mm/rev is:
(a) 3 seconds;
(b) 6 seconds;
(c) 10 seconds;
(d) 15 seconds.

3 A twist drill has its point thinned in order to:
(a) reduce the hole diameter;
(b) increase the rake angle;
(c) locate in the centre punch mark;
(d) reduce the axial (feed) pressure.

4 A reamer is used to correct the:
(a) size and roundness of a drilled hole;
(b) size and position of a drilled hole;
(c) finish and position of a drilled hole;
(d) finish and depth of a drilled hole.

5 Which of the following cutters should be used in a drilling machine

to recess the head of a cap screw?
(*a*) countersink bit;
(*b*) spot-facing cutter;
(*c*) counterbore;
(*d*) centre-drill.

Section B

6   The casting shown in Fig. 8.17 is to have a 25 mm diameter hole drilled
in the centre of boss 'B' and a 12 mm diameter hole drilled in the
centre of each of the bosses marked 'A'.
(*a*) The faces marked ⩒ have already been machined.
(*b*) The hole centres have already been marked out on the bosses.
Draw up an operation schedule for drilling the holes paying particular
attention to the: (i) method of work-holding and clamping; (ii) method
of locating the hole centres; (iii) limited feed pressure that can be
brought to bear on the unsupported casting under the bosses.

7   (*a*) Make a neat freehand sketch of a machine reamer, and on it
indicate the: (i) taper shank; (ii) bevel lead; (iii) flutes; (iv) body
length.
(*b*) Why do the flutes of a reamer have a *left-hand* helix?
(*c*) What is the advantage of a reamer having: (i) diametrically
opposed cutting edges; (ii) irregularly spaced cutting edges?
(*d*) Which faults in a drilled hole is a reamer used to correct?
(*e*) State *two* reasons why a reamer might leave a poor finish.

8   (*a*) Make a neat freehand sketch of a twist drill, and on it indicate
the: (i) clearance angle; (ii) rake angle; (iii) shank; (iv) chisel
point.
(*b*) With the aid of sketches explain why the diameter of a twist drill
and the web of a twist drill both taper towards the shank, but in
opposite directions.
(*c*) Sketch a typical application of a countersink cutter and of a
counter-bore cutter. Explain why the latter requires a pilot whilst
the former does not.

9   Sketch the feed mechanism of a bench drilling machine and explain
why it is referred to as a 'sensitive' feed mechanism.

Problems (the centre lathe)

Section A

1   The spindle speed for turning a 40 mm diameter bar at 30 m/min is:
(*a*) 239 rev/min;
(*b*) 377 rev/min;
(*c*) 418 rev/min;
(*d*) 424 rev/min.

2   The time taken for a cut 80 mm long at 320 rev/min if the feed rate
is 0.7 mm/rev is:
(*a*) 1.05 s;

(b) 21.4 s;

(c) 105 s;

(d) 168 s.

3 A straight-nose roughing tool is used for cutting:

(a) orthogonally;

(b) obliquely;

(c) profiles;

(d) square shoulders.

4 Concentricity is maintained by:

(a) cutting as few diameters as possible at each setting;

(b) re-setting in a self centring chuck (three-jaw);

(c) cutting as many diameters as possible at each setting;

(d) correct alignment of the tailstock.

Section B

5 The casting shown in Fig. 3.55 is to have a 25.05 + 0.05 mm diameter hole bored in the centre of boss 'B'.

1. The faces marked ✓ have already been machined.

2. The hole centre has already been marked out on the boss.

Draw up an operation schedule for boring the hole perpendicular to the face X on a centre lathe.

The schedule should include:

(a) the method of work holding and clamping;

(b) the method of locating the hole centre;

(c) balancing the face plate if necessary;

(d) the process of boring the hole;

(e) a method of checking the hole diameter whilst set up in the lathe.

6 Draw up an operation schedule for turning the bush shown in Fig. 3.59. To maintain concentricity the bush should be finished on a mandrel. Calculate a suitable cutting speed for turning the 15 mm diameter.

7 Name *three* basic techniques of taper turning used in the centre lathe. Tabulate the advantages and limitations of each of the techniques chosen.

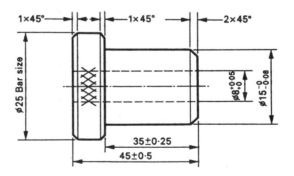

Fig. 3.59

8 (a) Describe, with the aid of sketches, two methods of setting work accurately in the four-jaw (independent) chuck.

(b) Sketch a travelling steady and explain how it is used to support long slender work.

4 Fastening and joining

4.1 Screwed fastenings

Various types of screwed fastenings are used where components must be assembled and dismantled regularly. Screwed fastenings are proportioned so that it requires the same force to make them fail in the following ways.

(a) By the head pulling off.

(b) By the thread stripping. (This assumes that the nut thickness is equal to at least the minor diameter of the thread.) For example, the nut thickness for an M12 thread is 10 millimetres, which compares with the minor diameter of an M12 thread which is 9.8 millimetres. If the bolt screws directly into a component the same proportions apply.

(c) By failure of the shank of the bolt or stud across the minor diameter (core diameter) of the thread.

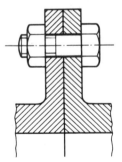

(a) Section through a bolted joint

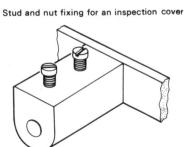

(b) Stud and nut fixing for an inspection cover

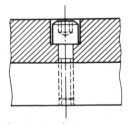

(c) Cap head socket screw

(d) Cheese head brass screws

Fig. 4.1 Use of screwed fastenings

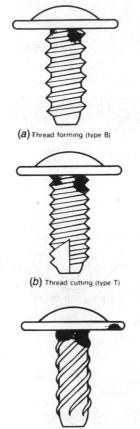

(*a*) Thread forming (type B)

(*b*) Thread cutting (type T)

(*c*) Drive (type U)

Fig. 4.2 Self-tapping screws

Screw thread systems were considered in Chapter 2, and Fig. 4.1 shows some typical applications of screwed fastenings.

(*a*) Figure 4.1(*a*) shows a section through a bolted joint. Note that the plain shank extends beyond the joint face so that no shearing load is carried across the thread.

(*b*) Figure 4.1(*b*) shows a stud and nut fixing for an inspection cover. This type of fixing is used where a joint is regularly dismantled. Most of the wear occurs on the stud and nut which can eventually be replaced cheaply and avoids wear falling on the threads of the expensive component to which the inspection cover is fixed.

(*c*) Figure 4.1(*c*) shows a cap-head socket screw. Although more expensive than ordinary hexagon head bolts, socket screws are made from high tensile alloy steel and are heat treated to make them very strong, tough, and wear resistant. They are widely used in the manufacture of machine tools, jigs and fixtures. The above example shows how the head is sunk into a counterbored hole to provide a flush surface.

(*d*) Figure 4.1(*d*) shows cheese-head screws. These are used for light duties where cost is important. In the example shown, they are being used for clamping an electrical cable into a terminal.

All the above fastenings are used in conjunction with a nut or with a thread which has been pre-cut into the component. *Self-tapping* screws are often used in soft materials such as plastics and also in thin sheet metal. As their name implies, such fastenings cut their own thread as they are driven into a tapping size hole. There are three basic types of self-tapping screw as shown in Fig. 4.2.

1. Thread-forming screws as shown in Fig. 4.2(*a*). These produce a thread by displacing the material around the pilot hole. They are used for joining non-ferrous sheet metal and sheet plastic.

2. Thread-cutting screws as shown in Fig. 4.2(*b*). These generate a thread by cutting into the wall of the pilot hole. They can be used for joining plastic and non-ferrous metal components and sheet steel.

3. Drive-screws as shown in Fig. 4.2(*c*). These are designed to be hammered or forced into a pilot hole. They can be used in a variety of materials and, although cheap to install, they can only be used once.

4.2 Locking devices

In order to prevent screwed fastenings slacking off due to vibration, various locking devices are employed. A selection of locking devices are shown in Fig. 4.3. It can be seen that they are divided into two categories; those where the locking action is positive, and those where the locking action is frictional. Positive locking devices are more time consuming and difficult to fit, but they are essential for critical joints where failure could cause serious accidents, as in the clutch and brake mechanisms of machine tools and road vehicles.

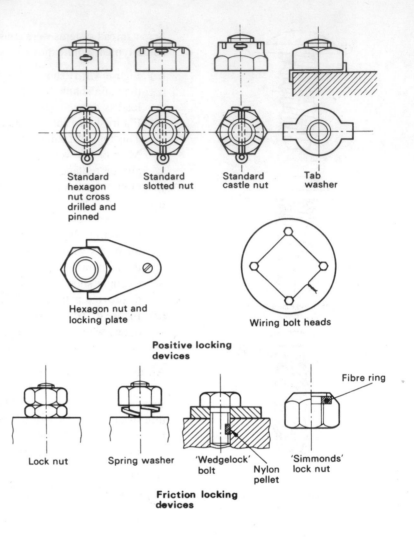

Standard hexagon nut cross drilled and pinned

Standard slotted nut

Standard castle nut

Tab washer

Hexagon nut and locking plate

Wiring bolt heads

Positive locking devices

Lock nut

Spring washer

'Wedgelock' bolt

Nylon pellet

Fibre ring

'Simmonds' lock nut

Friction locking devices

Fig. 4.3 Locking devices

4.3 Tightening screwed fastenings

To obtain the optimum joint strength a screwed fastening must be tightened up correctly. Spanners are carefully proportioned so that their length enables a man of average strength to fully tighten the fastening correctly but not over-tighten it. Extending a spanner to exert greater force on the fastening can result in the fastening being stressed beyond the elastic limit so that it takes on a permanent set and is not only severely weakened but loses its elasticity. The elasticity or 'spring back' in a screwed fastening when it is tightened provides the clamping force which holds the joint faces tightly together. If the elasticity is destroyed by incorrect use then fastening becomes useless and the joint is no longer secure.

In order to ensure that critical screwed fastenings are tightened correctly a *torque spanner* should be used. Figure 4.4 shows such a spanner. These fall into two categories. One type uses a pre-set slipping clutch which can be set to slip when the nut is correctly tightened and prevents over-tightening. The other, as shown in Fig. 4.4, has

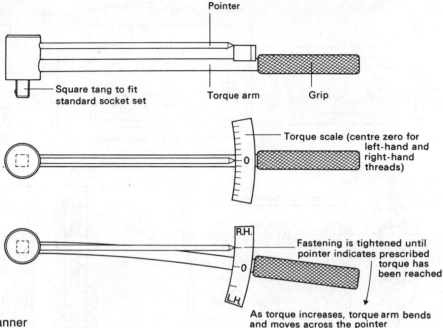

Fig. 4.4   Typical torque spanner

a spring loaded handle. A pointer indicates on a scale the force being exerted on the nut. This latter type relies heavily upon the skill of the fitter for its correct functioning. However it is much cheaper and simpler in construction.

4.4   Riveted joints

Riveting is a method of making permanent joints. This process consists of drilling or punching holes in the plates to be riveted, inserting the rivet, then closing it by applying a compression force which causes the rivet to swell and fill the hole and, at the same time, forms a head on the end of the rivet to secure it in place. Note, the strength of a rivet is in its shank and not in its head. Therefore a riveted joint should be so designed that the major forces on the joint place the rivet in *shear* and NOT in *tension*. A variety of riveted joints are used in construction and fabrication work, some of the more common being shown in Fig. 4.5.

Single riveted lap joint

This is the simplest of all riveted joints and is extensively used for joining both thick and thin plates. The plates to be joined are overlapped a short distance and a single row of rivets, spaced to give the required strength, are inserted along the middle of the lap to complete the joint.

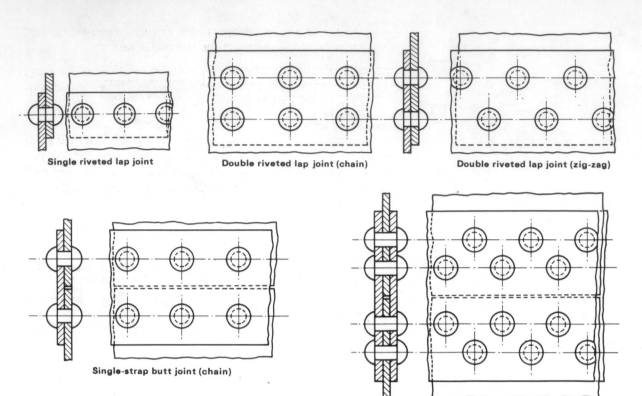

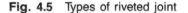

Fig. 4.5 Types of riveted joint

Double riveted lap joint

This is a lap joint with two rows of rivets. Sufficient overlap must be provided in order to accommodate the two rows of rivets and leave enough material between the rows to give a joint of the required strength. The two rows of rivets may be arranged in a square formation known as 'chain riveting', or the rivets may be arranged diagonally to form triangles, called 'zig-zag riveting'.

Single-strap butt joint

To make a riveted butt joint it is necessary to use a separate piece of metal called a strap to join the two component edges together. The strap is usually the same type of material and the same thickness as the plates being joined.

Double-strap butt joint

When two straps (also called a strap and a cover-plate) are used, one on each side of a butt joint, the joint is known as a 'double-strap butt joint'. No matter whether a single or a double strap is used for riveted butt joints the arrangements of the rivets may be:

(a) single riveted (i.e. one row of rivets on each side of the butt),
(b) double, triple or quadruple riveted (i.e. two, three, or four rows of rivets each side of the butt, arranged in 'chain' or in 'zig-zag' formation).

The standard types of rivet head are shown in Fig. 4.6. Flat-head rivets are used for most general sheet metal work where the metal is thin and little strength is required. When flat head rivets are coated with tin to prevent corrosion and make them easier to soft-solder they are

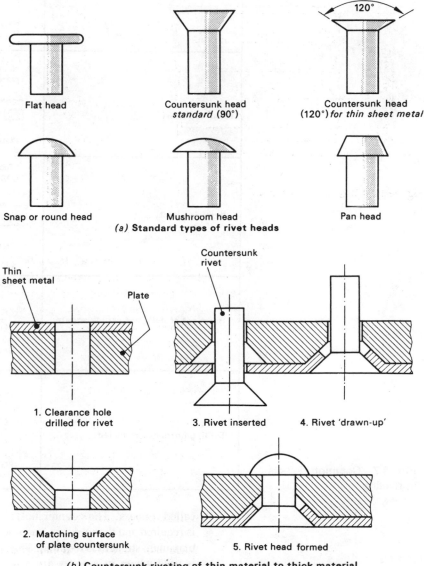

Fig. 4.6 Rivet heads and applications

CAUSE OF RIVETING DEFECT	RESULTANT EFFECT
Insufficient hole clearance	Rivet not completely 'drawn through' Not enough shank protruding to form head  Original head of rivet 'stands proud', the formed head is weak and mis-shapened
Hole too large for rivet	Hole not filled  Rivet tends to bend and deform. Head weak and poorly shaped
Rivet too short	Not enough shank protruding to produce a correctly shaped head 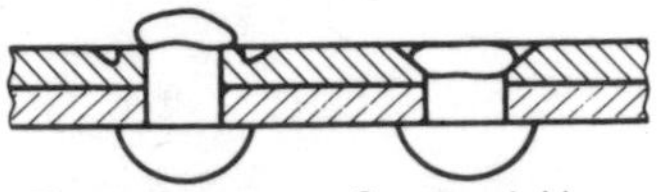 Plate surface damaged Countersinking not completely filled
Rivet too long	Too much shank protruding to form required head  'Flash' formed around head (Jockey cap) Countersinking over-filled
Rivet set or dolly not struck square	Badly shaped head - off centre 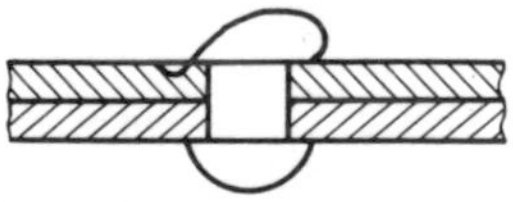 Sheet damaged by riveting tool
Drilling burrs not removed	Not enough shank protruding to form the correct size head 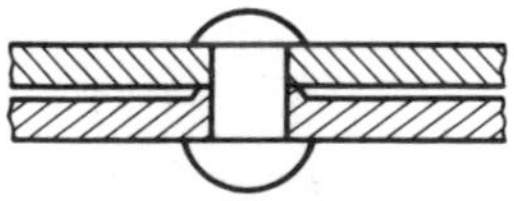 Plates or sheets not closed together. Unequal heads

Fig. 4.7 Common defects in riveting

called 'tinners'. The countersunk rivet is used where a flush surface is required and the round or 'snap-head' rivet is commonly used where maximum strength is required. The mushroom head or 'knobbled' rivet is used where it is necessary to curtail the height of the rivet. Pan head rivets are very strong and are used for girder work and heavy

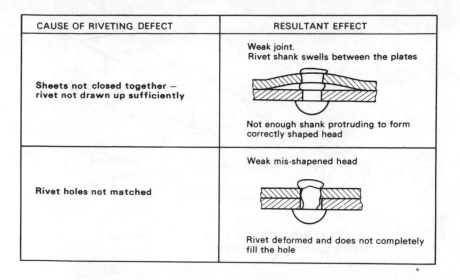

CAUSE OF RIVETING DEFECT	RESULTANT EFFECT
Sheets not closed together — rivet not drawn up sufficiently	Weak joint. Rivet shank swells between the plates Not enough shank protruding to form correctly shaped head
Rivet holes not matched	Weak mis-shapened head Rivet deformed and does not completely fill the hole

Fig. 4.7 *(continued)*

constructional engineering as an alternative to the snap-head rivet. Generally, the rivet should be made from the same metal as the components being joined. Some of the more common defects associated with riveted joints are shown in Fig. 4.7.

Hollow components present special problems when riveting because of the difficulty in getting a hold-up or 'dolly' behind the rivet whilst it is being headed. Cylindrical components can be riveted using a *bench mandrel* as shown in Fig. 4.8(*a*), whilst Fig. 4.8(*b*) shows the principle of *'pop' riveting* which can be used for box section components where it is impossible to insert any sort of hold-up.

4.5 Soft soldering

The process of soft soldering involves the use of a suitable low melting temperature alloy of tin and lead which is 'bonded' by the application of heat and a suitable 'flux' to an unmelted parent metal. Therefore, soft solders must have a lower melting temperature than the metals they are joining. The compositions of some typical soft solders were given in Table 1.14, p. 24.

It is an essential feature of a soldered joint that each of the joint surfaces is 'tinned' by a film of solder and that these two films of solder are made to 'fuse' with the solder filling the space between them. In a correctly soldered joint, examination under a microscope will show that the action of tinning metals such as copper, brass and sheet steel causes the solder and the parent metal to react together to form an *intermetallic compound* which acts as a 'key' for the bulk of the solder in the joint. This intermetallic compound layer will continue to grow in thickness for as long as the joint is kept at the soldering temperature. Solder can never be completely wiped or drained from the surface of the metal once tinning has occurred. Therefore the surface of metal remains permanently 'wetted' or 'tinned' by a film of solder which cannot be mechanically prised or scraped off to leave the metal surface in its original state.

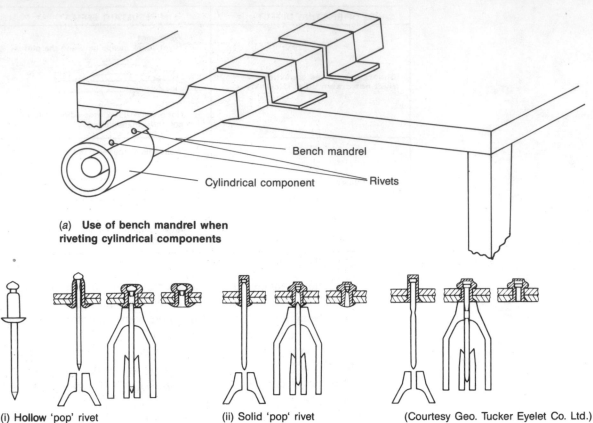

(a) **Use of bench mandrel when riveting cylindrical components**

(i) Hollow 'pop' rivet   (ii) Solid 'pop' rivet   (Courtesy Geo. Tucker Eyelet Co. Ltd.)

Fig. 4.8 Riveting hollow components

(b) **The 'pop' rivet system**

The tinning action of the solder cannot take place unless the two surfaces to be joined are physically and chemically clean. This is achieved by *fluxing* the joint faces after scouring them to remove any dirt and grease. The soldering flux then removes the oxide film and prevents it reforming whilst soldering is taking place. The requirements of a flux for soft soldering are that it must:

(a) clean the oxide film from the surface to be soldered;
(b) prevent the oxide film reforming during the soldering process;
(c) 'wet' the surfaces being joined so that the solder runs evenly over the surface and does not roll up into globules;
(d) be easily displaced by the molten solder so that a metal to metal contact is achieved.

Figure 4.9 shows the essential functions of a soldering flux whilst tinning a metal surface. Fluxes used for soft soldering operations may be classified as 'active' and 'inactive' or 'passive'.

Active fluxes such as Baker's Fluid (acidified zinc chloride solution) quickly dissolve the oxide film and prevent oxidation. They etch the surface being soldered and ensure good wetting and bonding.

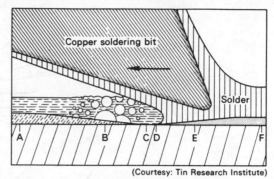

A molten solder is said to 'wet' when it leaves a continuous permanent film on the surface of the parent metal instead of rolling over it.

(Courtesy: Tin Research Institute)

Diagrammatic representation of the displacement of flux by molten solder.

A Flux solution lying above oxidised metal surface.
B Boiling flux solution removing the film of oxide (e.g. as chloride).
C Bare metal in contact with fused flux.
D Liquid solder displacing fused flux.
E Tin reacting with the basis metal to form compound.
F Solder solidifying.

Fig. 4.9 The essential functions of a soldering flux

Unfortunately, all active fluxes leave a corrosive residue along the edges of the joint after soldering. These apparently dry residues are hygroscopic, that is, they absorb water from the atmosphere. As soon as the residue becomes moist it attacks the metal being joined and causes severe corrosion. If an active flux has to be used, it is essential that any residue is washed off the component immediately after soldering and the joint treated with a rust inhibitor.

Passive fluxes such as resin are used for those applications where it is not possible to remove the corrosive residue by washing, for example, electrical terminations. Unfortunately passive fluxes do not clean the oxide film from the joint faces to any appreciable extent, they only prevent the film reforming during the soldering process. Thus the initial mechanical cleaning of the joint faces must be very thorough.

Figure 4.10 shows the stages in making a soldered joint by the technique called 'sweating'.

(a) A thin film of flux is applied along the edges of the pieces of metal being joined after they have been cleaned.

(b) The materials to be joined are placed on a wooden block or similar material which is non-conductive of heat. A tinned and loaded soldering iron is drawn slowly along the ends of the work as shown. As heat energy is transferred from the 'bit' to the work the temperature of the work is raised. When it reaches the melting point of the solder being used, solder is transferred from the bit to the work, and the joint surfaces become 'tinned'.

(c) The work is then placed together with the tinned surfaces in contact and are held by pressing down on them with a stick or the tang of an old file. A heated soldering iron is then placed on one end of the joint ensuring maximum surface contact is made between the 'bit' and the work. As the solder from the pre-tinned

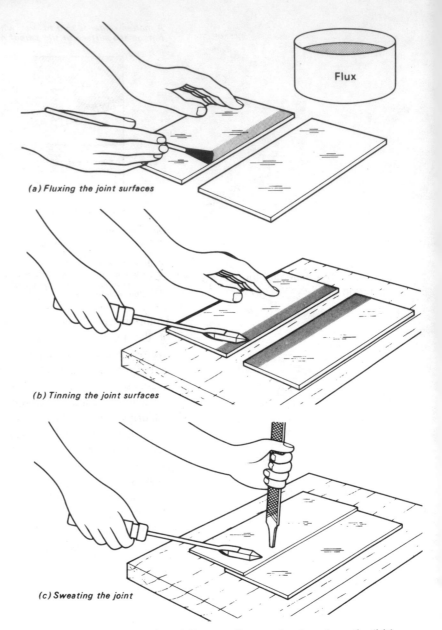

(a) Fluxing the joint surfaces

(b) Tinning the joint surfaces

(c) Sweating the joint

Fig. 4.10   Soft soldering

surfaces starts to melt and flow out from under the edges, the 'bit' is slowly drawn along the joint followed by the hold-down stick to ensure that the solder film is kept to a minimum. The success of sweating depends upon having an adequate and constant supply of heat energy. The 'bit' of the soldering iron must be as large as possible for the joint being made and, as previously stated, the work must be supported so that there is minimum heat loss. Note that a satisfactory join will not be achieved unless the material being joined is itself raised to just above the melting temperature of solder during tinning and during the final sweating.

4.6   Hard soldering and brazing

Hard soldering is a general term used to cover *brazing* and *silver soldering*. In these processes, as in soft soldering, melting or fusion of the parent metals to be joined does not take place, which means that a 'filler' material must be used which has a lower melting temperature than the 'parent metal' of the workpiece.

There are a number of alloys other than the tin-lead (soft solder) alloys which can be used as a solder. They do not possess the very low fusion (melting) temperatures of the soft solders, but they are much stronger which makes them more suitable for certain jobs. For example, bicycle frames are traditionally made by *brazing* steel tubes into the corner brackets. In this case the solder is made from a grade of *brass* (hence the name *brazing*), which usually consists of 60% copper and 40% zinc. This alloy melts at 850 °C, which is much higher than soft solder. Such high melting point solders are called *hard solders*, and hard soldered or brazed joints are not only much stronger than soft soldered joints, but their strength is retained at much higher temperatures.

Hard soldering (brazing) can be defined as:

> *a process of joining metals in which a molten filler metal is drawn by capillary attraction into the space between closely adjacent surfaces of the parts to be joined.*

In general, the melting temperature of the filler material lies above 500 °C but below the melting temperature of the parent metal being joined. In this respect hard soldering lies between soft soldering and welding. The success of all hard soldering operations depends upon the following general conditions.

(a) Selection of a suitable filler alloy which has a melting range appreciably lower than that of the parent metals to be joined.

(b) Thorough cleanliness of the surfaces to be hard soldered.

(c) Complete removal of the oxide film from the joint surfaces before and during hard soldering by a suitable flux.

(d) Complete 'wetting' of the joint surfaces by the molten filler alloy. When a surface is 'wetted' by a liquid, a continuous film of the liquid remains on that surface after draining. This condition is essential for hard soldering and the flux, having removed the oxide film, must completely 'wet' the joint surfaces. This wetting action by the flux assists the spreading and feeding of the molten filler alloy into the joint by capillary attraction. This ensures a completely filled joint.

(e) Since the molten filler alloy is drawn into the joint by capillary attraction, the space between the joint faces must be kept to a minimum and kept *constant*. Any local increase in the gap can present a barrier to the feeding of the filler alloy and will prevent the joint being uniformly filled causing a severe loss of strength.

(f) Melting the filler alloy alone is not sufficient to produce a sound joint. The parent metal must itself be raised to the brazing temperature so that the filler alloy melts on coming into contact with the joint surfaces.

Unlike welding, dissimilar metals and alloys may be successfully joined by hard soldering. For example: copper to brass; brass to steel; mild steel to malleable cast iron, etc. The groups of filler materials most widely used for hard soldering are as follows.

Silver solders

These are more expensive than the copper based alloys because they contain a high proportion of the precious metal *silver*. However they offer the advantages of producing very strong and ductile joints at much lower temperatures than when using the copper based alloys. Silver solders are also very free flowing at the process temperature which is sufficiently low to have little effect on the parent metal. The use of silver solder alloys speeds up the process and reduces the need for 'finishing' since neat precision joints can be made. Borax is not a suitable flux for silver soldering and a proprietary flux to match the grade of solder should be used.

Brazing alloys containing phosphorus

Filler metal and alloys containing *phosphorus* are usually referred to as 'self-fluxing alloys'. These alloys contain silver, phosphorus and copper for low temperature work, or just copper and phosphorus where lower cost and greater strength is required. These filler alloys are used to braze *copper components* in air without the use of a separate flux. They are only effective when melted in an oxidising atmosphere since the products of oxidation form a molten compound which is self-fluxing. Brazing alloys containing phosphorus should only be used when brazing components made from copper and copper alloys excepting those containing more than 10% nickel. They must never be used to braze nickel, nickel-based alloys and ferrous metals since they react with these metals to produce weak joints.

Brazing spelters

The oldest and best known method of brazing involves the use of brazing brasses ('spelters') as the filler alloy and *borax* as the flux. These alloys melt at much higher temperatures than the silver solders and the phosphorus self-fluxing alloys, but they make stronger joints. Increasing the zinc content of the spelter increases its melting range and enables the process to be carried out at lower temperatures so that, with care, even 60/40 brass can be brazed. However increasing the zinc content also reduces the strength of the joint. Some of the zinc vaporises off so that the final joint material has a lower zinc content than the initial specification of the spelter would imply.

Aluminium brazing

For aluminium brazing, the filler material is an aluminium alloy of suitable composition and having a melting point below that of the parent metal. Borax based fluxes are unsuitable for brazing aluminium and its alloys, but many proprietary fluxes are available. These are, basically, mixtures of alkali metal chlorides and fluorides. A typical flux contains chlorides of sodium, potassium, and lithium. The melting temperatures of the parent metals, which can be aluminium brazed, range between 590 °C and 660 °C. Extreme care must be taken when aluminium brazing because of the very small margin between the process temperature and the temperature at which the parent metal starts to melt.

The simplest and most versatile hard soldering process is traditional *flame brazing*. It is equally applicable to silver soldering and to brazing, the only difference being in the materials used and the process temperature. Flame brazing may be used to fabricate almost any assembly and particularly where the joint area is small in relation to the bulk of the assembly. Gas torches using oxy-propane, air-propane, or air-methane (natural gas) mixtures are most commonly used. Propane is preferable to natural gas as it burns at a higher temperature. Natural gas is only really suitable for the lower melting point filler alloys. The flux is made into a stiff paste with water and brushed round the joint. The assembled components are then brought up to temperature with the gas torch as shown in Fig. 4.11, and the flux melts and runs into the joint. The filler alloy, in strip form, is then brought into contact with the work, whereupon it melts and runs into the joint. Note that it is the heat of the work which must melt the filler alloy, *not* the flame of the torch. It is usual to coat the silver solder or brazing spelter with flux by dipping it into the paste before melting it into the joint. Figure 4.12 shows various types of brazed joints designed to provide maximum area of bonding between the components being joined.

4.7 Fusion welding

In the soldering and brazing processes described so far, the joints are formed by a thin film of metal that has a lower melting point and inferior strength to the metals being joined. In *fusion welding* any additional metal added to the joint has a similar composition and strength to the metals being joined. Figure 4.13 shows the principle of joining two pieces of metal by fusion welding where not only the filler metal but also the edges of the components are being melted. The molten metals fuse together and, when solid, form a homogeneous joint whose strength is equal to the metal of the components being joined.

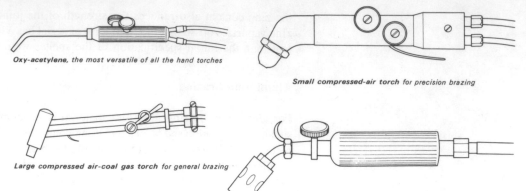

Oxy-acetylene, the most versatile of all the hand torches

Small compressed-air torch for precision brazing

Large compressed air-coal gas torch for general brazing

Air-propane torch for low temperature brazing

(a) Typical hand torches used for brazing

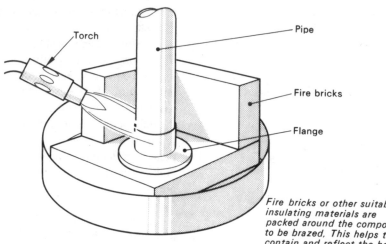

Fire bricks or other suitable insulating materials are packed around the component to be brazed. This helps to contain and reflect the heat supplied by the torch.

Fig. 4.11 Flame brazing   *(b)* Hand torch in use with brazing hearth

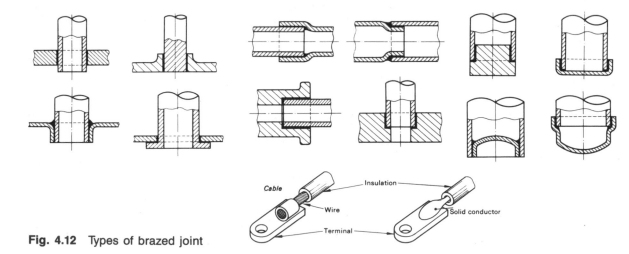

Fig. 4.12 Types of brazed joint

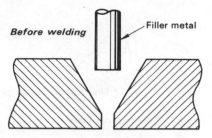

Before welding — Filler metal

SINGLE VEE BUTT requires extra metal

After welding

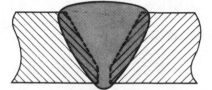

*The edges of vee are melted and fused
together with the molten filler metal*

Fig. 4.13   Fusion welding

Oxy-acetylene welding

In this process the heat source is a mixture of oxygen and acetylene gases burning to produce a flame whose temperature can reach 3250 °C, and this is above the melting point of most metals. Figure 4.14 shows a typical set of gas welding equipment. Since the gases are stored under very high pressures and form highly flammable and even explosive mixtures the equipment must be handled with great care. This equipment must only be used by persons who have been fully instructed in the operating and safety procedures recommended by the Home Office and by the equipment suppliers. Figure 4.15 shows the two basic techniques for fusion welding using an oxy-acetylene torch. No flux is required when welding ferrous metals as the products of combustion from the burnt gases protect the molten weld pool from atmospheric oxygen. The three types of flame produced by a welding torch are shown in Fig. 4.16, and the *neutral flame* is the one normally used.

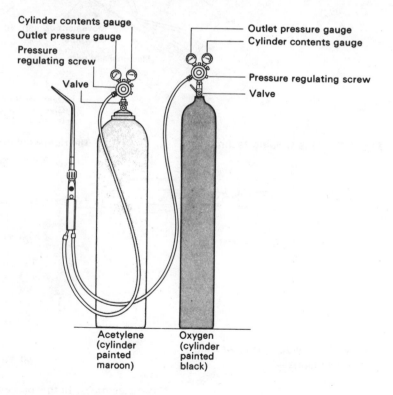

Cylinder contents gauge
Outlet pressure gauge
Pressure
regulating screw
Valve

Outlet pressure gauge
Cylinder contents gauge

Pressure regulating screw
Valve

Acetylene
(cylinder
painted
maroon)

Oxygen
(cylinder
painted
black)

Fig. 4.14   Oxy-acetylene welding equipment

Metallic arc welding

This is a fusion welding process where the energy required to melt the edges of the components and the filler rod is provided by an electric arc. The *arc* is the name given to a prolonged spark struck between

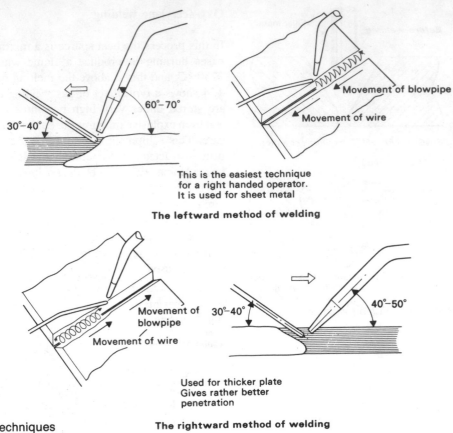

This is the easiest technique
for a right handed operator.
It is used for sheet metal

The leftward method of welding

Used for thicker plate
Gives rather better
penetration

The rightward method of welding

Fig. 4.15 Gas welding techniques

(a) **The neutral flame**

(b) **The oxidising flame**

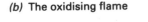

Fig. 4.16 Oxy-acetylene welding
flame conditions

(c) **The carburising flame**

two electrodes. In this process the filler rod forms one electrode and
the workpiece forms the other electrode. The filler rod/electrode is
coated with a flux which shields the joint from atmospheric oxygen
at the very high temperatures involved. (Average arc temperature is
about 6000 °C.) Figure 4.17 compares the principles of gas and arc
welding. A transformer is used to reduce the mains voltage to a safe,
low voltage, heavy current supply suitable for welding. As with gas

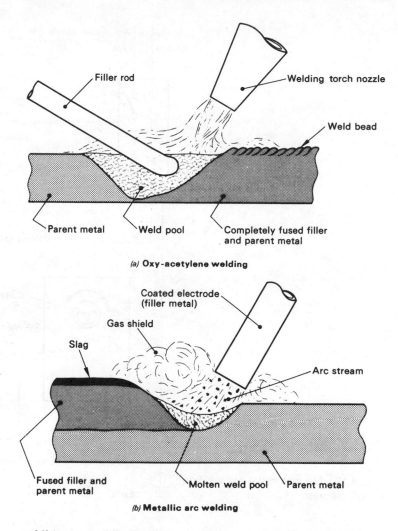

Fig. 4.17 Comparison of oxy-acetylene and metallic arc welding

(a) **Oxy-acetylene welding**

(b) **Metallic arc welding**

welding, arc welding equipment must not be used by untrained persons, except under the closest supervision. The dangers associated with arc welding arise from the very high temperatures and heavy currents involved, together with the risk of electrocution if the equipment is not correctly installed, maintained, and used. Figure 4.18 shows the general arrangement of a metallic welding installation.

4.8 Use of adhesives

Traditionally, adhesives fell into two categories:

1. *Glues*: These are made from the bones, hooves and horns of animals and the bones of fishes. Derivatives of milk and blood are also used. Glues are largely used for jointing wood and are used in the furniture and toy manufacturing industries.
2. *Gums*: These are made from vegetable matter, resins and rubbers being extracted from trees and the starches being extracted from the by-products of flour milling.

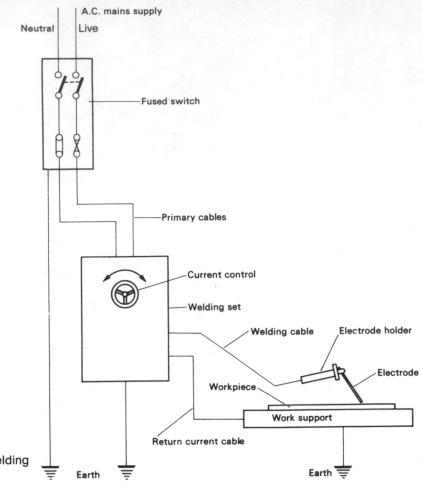

Fig. 4.18 Manual metal-arc welding circuit diagram

Although natural glues and gums are still widely used for low strength, non-toxic applications, they are being increasingly supplanted by high strength synthetic adhesives developed by the polymer (plastics) industries. Table 4.1 lists some of the more important advantages and limitations of modern high-strength, synthetic adhesives compared with the mechanical and thermal joining techniques discussed earlier in this chapter.

Figure 4.19(*a*) shows a typical bonded joint and explains the terminology used for the various features of the joint. The strength of the bond depends upon the following two factors.

(*a*) *Adhesion* is the ability of a bonding material (adhesive) to stick (adhere) to the materials being joined (adherends). There are two ways in which the bond can occur and these are shown in Fig. 4.19(*b*).

(*b*) *Cohesion* is the ability of the adhesive film itself to resist the applied forces acting on the joint.

Table 4.1 Advantages and limitations of bonded joints

Advantages
1. The ability to join dissimilar materials, and materials of widely different thicknesses
2. The ability to join components of difficult shape that would restrict the application of welding or riveting equipment
3. Smooth finish to the joint which will be free from voids and protrusions such as weld beads, rivet and bolt heads, etc.
4. Uniform distribution of stress over entire area of joint. This reduces the chances of the joint failing in fatigue
5. Elastic properties of many adhesives allow for flexibility in the joint and give it vibration damping characteristics
6. The ability to electrically insulate the adherends and prevent corrosion due to galvanic action between dissimilar metals
7. The joint will be sealed against moisture and gases
8. Heat-sensitive materials can be joined

Limitations
1. The bonding process is more complex than mechanical and thermal processes, i.e. the need for surface preparation, temperature and humidity control of the working atmosphere, ventilation and health problems caused by the adhesives and their solvents. The length of time that the assembly must be jigged up whilst setting (curing) takes place
2. Inspection of the joint is difficult
3. Joint design is more critical than for many mechanical and thermal processes
4. Incompatibility with the adherends. The adhesive itself may corrode the materials it is joining
5. Degradation of the joint when subject to high and low temperatures, chemical atmospheres, etc.
6. Creep under sustained loads

Figure 4.19(c) shows three ways in which a bonded joint may fail under load. These failures can be prevented by careful design of the joint, correct selection of the adhesive, careful preparation of the joint surfaces and control of the working environment. (Cleanliness, temperature and humidity.)

No matter how effective the adhesive is and how carefully it is applied, the joint will be a failure if it is not correctly designed and executed. It is bad practice to apply adhesive to a joint which was originally proportioned for bolting, riveting, and/or welding. The joint must be proportioned to exploit the special properties of adhesives. Most adhesives are relatively strong in tension and shear, but weak in cleavage and peel. These terms are explained in Fig. 4.20. The adhesives must 'wet' the joint surfaces thoroughly, otherwise voids will occur and the bonded area will be considerably less than the theoretical maximum: this will weaken the joint considerably. Figure 4.21 shows the effect of wetting on the adhesive film.

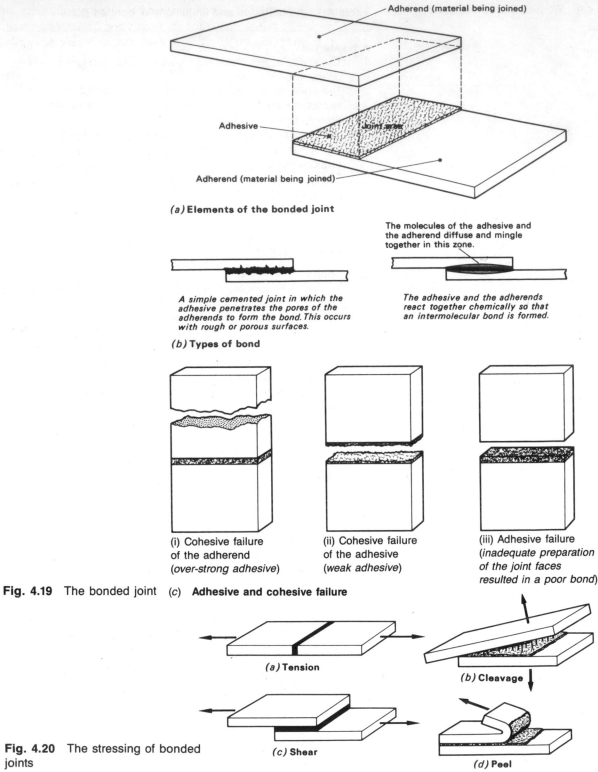

Adherend (material being joined)

Adhesive

Joint area

Adherend (material being joined)

(a) **Elements of the bonded joint**

A simple cemented joint in which the adhesive penetrates the pores of the adherends to form the bond. This occurs with rough or porous surfaces.

The molecules of the adhesive and the adherend diffuse and mingle together in this zone.

The adhesive and the adherends react together chemically so that an intermolecular bond is formed.

(b) **Types of bond**

(i) Cohesive failure of the adherend *(over-strong adhesive)*

(ii) Cohesive failure of the adhesive *(weak adhesive)*

(iii) Adhesive failure *(inadequate preparation of the joint faces resulted in a poor bond)*

Fig. 4.19 The bonded joint   *(c)* **Adhesive and cohesive failure**

(a) **Tension**

(b) **Cleavage**

(c) **Shear**

(d) **Peel**

Fig. 4.20 The stressing of bonded joints

4.9 Thermoplastic adhesives

Thermoplastic materials were introduced in Section 1.18. They are materials which soften when they are heated and harden again when cooled. The adhesives derived from these materials may be applied in three ways.

*An adhesive with a **poor wetting action** does not spread evenly over the joint area. This reduces the effective area and weakens the joint.*

*An adhesive with a **good wetting action** will flow evenly over the entire joint area. This ensures a sound joint of maximum strength.*

Fig. 4.21 Wetting capacity of an adhesive

1 Heat activated

The adhesive is softened by heating until it is fluid enough to spread freely over the joint surfaces. These are then brought into contact and pressure is applied until the adhesive has cooled to room temperature and set. The traditional 'glues' are naturally-occurring heat-activated thermoplastic adhesives.

2 Solvent activated

Instead of heat, the adhesive is softened by a suitable solvent. The dissolved adhesive is applied to the joint and a bond is achieved by the solvent evaporating. The 'balsa cement' used by aero-modellers is a solvent activated adhesive consisting of a cellulose material dissolved in highly volatile acetone. Because evaporation is essential to the setting of the adhesive, a sound bond is almost impossible to achieve at the centre of a large joint area as shown in Fig. 4.22. This is particularly the case when joining non-absorbent materials.

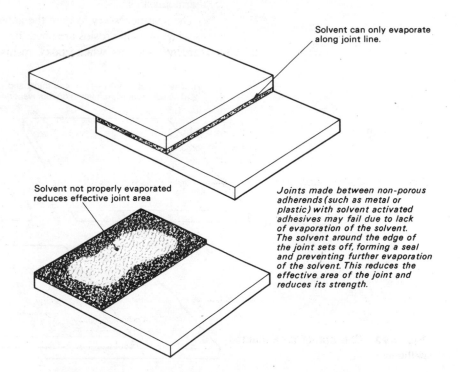

Solvent can only evaporate along joint line.

Solvent not properly evaporated reduces effective joint area

Joints made between non-porous adherends (such as metal or plastic) with solvent activated adhesives may fail due to lack of evaporation of the solvent. The solvent around the edge of the joint sets off, forming a seal and preventing further evaporation of the solvent. This reduces the effective area of the joint and reduces its strength.

Fig. 4.22 Solvent activated adhesive fault

3 Impact adhesives

These are solvent activated adhesives which are spread separately on the two joint faces and left to dry by evaporation. When dry, the treated faces are brought together whereupon they instantly bond together by intermolecular attraction. This enables non-absorbent materials to be successfully joined over a large contact area as evaporation takes place prior to the joint faces being brought into contact. Figure 4.23 shows the steps in making an impact joint.

Thermoplastic adhesives are based upon synthetic materials such as polyamides, vinyl and acrylic polymers and cellulose derivatives. They are also based upon natural materials such as resins, shellac, mineral waxes, and rubber. They are not so strong as thermosetting plastics but, being generally more flexible, they are more suitable for joining non-rigid materials. Unfortunately they are heat sensitive and lose their strength rapidly as the ambient temperature rises. For example, the natural glues become liquid at the temperature of boiling water.

4.10 Thermosetting adhesives

Thermo-setting plastic materials were introduced in Section 1.18. These are materials which depend upon heat to make them set. The setting (curing) process causes chemical changes to take place within the adhesive, and once set (cured) they cannot be softened again by the re-application of heat. This makes them less heat sensitive than the thermoplastics.

The heat necessary to cure the adhesive can be applied externally as when phenolic resins are used, or internally by adding a chemical hardener such as when epoxy resins are used. The hardener is a

1. The impact adhesive is spread thinly and evenly on both joint surfaces.
2. The adhesive is then left to dry be evaporation. This avoids the problem in Fig. 4.22.

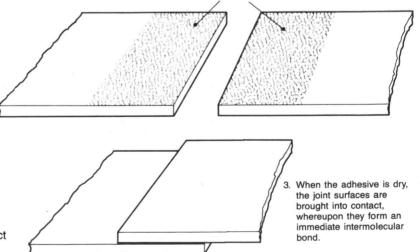

3. When the adhesive is dry, the joint surfaces are brought into contact, whereupon they form an immediate intermolecular bond.

Fig. 4.23 The use of an impact adhesive

chemical which reacts with the adhesive to generate heat (exothermic reaction) and cause polymerisation (curing). Since the setting process is a chemical reaction and is not dependent upon evaporation, the area of the joint does not affect the setting process. Thermosetting adhesives are very strong and are used in making structural joints in high strength materials such as metals. Joints in the body shells of motor cars and stressed components of aircraft are becoming increasingly dependent upon adhesives in place of spot welding and riveting. The stresses are more uniformly transmitted from one component of the joint to the next, and the joints are sealed against corrosion. Furthermore, the relatively low temperatures involved in adhesive bonding do not affect the grain structure and properties of the parent metal as does welding and brazing. Thermosetting adhesives tend to be rigid when cured and, therefore are unsuitable for joining flexible materials.

4.11 Safety in the use of adhesives

One great advantage of natural gums and glues is that they are non-toxic and are not particularly flammable. Therefore they are widely used in the labelling and packaging of foodstuffs. Unfortunately most synthetic adhesives and their solvents, hardeners, catalysts, etc., are highly toxic and some also have narcotic effects. In addition the solvents used in thermoplastic and impact adhesives are highly volatile and flammable. Therefore, synthetic adhesives together with their solvents, hardeners, and catalysts must be stored and used only in well ventilated conditions and the working area must be declared a *no-smoking* zone.

The health hazards presented by these materials range from dermatitis and sensitisation of the skin, to permanent damage of the brain and internal organs if inhaled or accidently swallowed.

Precautions

(a) Use only in well ventilated areas.
(b) Wear protective clothing appropriate to the process, no matter how inconvenient.
(c) Use a barrier cream.
(d) After use, wash thoroughly in soap and water; do *not* use solvents.
(e) Do not smoke in the presence of solvents.
(f) Obey all safety procedures and codes of practice.

Problems

Section A

1 For maximum strength a riveted joint should be arranged so that the rivet is in:
(a) shear; (b) tension; (c) compression; (d) no way stressed.

2 The soldering flux known as 'killed spirits' is:
  (*a*) non-corrosive;
  (*b*) inactive;
  (*c*) active;
  (*d*) a paste.

3 Self-secured joints are made by:
  (*a*) sweating the edges of sheet metal together;
  (*b*) interlocking the edges of folded sheet metal;
  (*c*) using an impact adhesive;
  (*d*) riveting the edges of sheet metal together.

4 Locking devices are used to prevent screwed fastenings:
  (*a*) being over tightened;
  (*b*) being stolen;
  (*c*) working loose;
  (*d*) being dismantled.

5 Most adhesives are relatively strong in:
  (*a*) tension and shear;
  (*b*) cleavage and peel;
  (*c*) tension and cleavage;
  (*d*) shear and peel.

Section B

6 Describe, with the aid of sketches, how the following joints are made:
  (*a*) a grooved seam between two pieces of tin-plate;
  (*b*) a bolted joint between two 12 mm thick plates in which one of the plates has an M10 $\times$ 1.5 tapped hole;
  (*c*) a single-strap butt joint, doubled riveted with the rivets in 'zig-zag' formation.

7 (*a*) Explain what the main purposes are of a flux when soft soldering.
  (*b*) Give an example of an active flux and an example of a passive flux. Explain what the main differences are between these two types of flux and give an example of a typical application of each type.
  (*c*) Explain what the main differences are between plumber's solder and tinman's solder and give an example of a typical application of each type.

8 (*a*) Describe the essential differences between hard soldering (brazing) and soft soldering, and describe the four basic conditions upon which a successful brazed joint depends.
  (*b*) Explain, with the aid of sketches, how a collar may be brazed on to a shaft using electric induction heating.

9 (*a*) Sketch a section through a corner joint between two pieces of sheet metal that has been designed for adhesive bonding.
  (*b*) Describe how the joint surfaces should be prepared, and what precautions should be taken to ensure a sound joint.
  (*c*) Select a suitable adhesive for a high strength joint between the two pieces of metal and give reasons for your choice.

10 Describe under what circumstances the following joints would be used:

(*a*) a screwed joint;

(*b*) a riveted joint;

(*c*) a self-secured joint;

(*d*) a soft soldered joint;

(*e*) a hard soldered joint;

(*f*) an adhesive bonded joint.

Your answer should take into consideration the following factors: the conditions under which the joint has to function, the properties of the materials being joined and the effectiveness of the joint.

5 Working in plastics

5.1 The joining of plastics

Depending upon their composition, plastic materials may be joined by mechanical methods such as riveting, the use of nuts and bolts, or by self-tapping screws. Screwed connections are only used when a semi-permanent joint is required. However these joining techniques tend to concentrate the load at the points where the connections are made, and plastics are better joined by processes which avoid stress concentrations and spread the load over a larger area using one of the following techniques.

5.2 Heat welding

Heat welding can only be used to join *thermoplastic materials* since only these plastic materials soften upon heating. A distinction is made in the joining of plastics between welding and sealing. The term *sealing* is reserved for the thermal joining of thin films and foils (plastic bags containing food such as potato crisps), whereas the term *welding* is applied to the joining of relatively heavy gauge (thick) sheet plastic components. The technique of welding sheet plastic is similar to that of oxy-acetylene welding in that a heat source and filler rod are used and the edges of the parent material and the filler material are softened and merge together. Obviously, an oxy-acetylene torch cannot be used when welding plastics and a more gentle means of heating is required. The low thermal conductivity and softening temperatures of plastic materials necessitates the use of a low welding temperature so that the heat can penetrate to the body of the plastic before the surface overheats and degrades.

Heat is normally applied to the joint by a welding 'gun'. Figure 5.1(*a*) shows an electrically heated welding gun. This is similar in construction to a hair dryer. A fan driven by an electric motor blows air past an electric resistance heater and the heated air is expelled through a jet so that it can be directed onto the weld zone. Figure 5.1(*b*) shows a gas heated gun using acetylene or propane gas in a 'bunsen' type burner. Instead of air this 'gun' is directing heated nitrogen gas onto the weld zone to protect the hot plastic from atmospheric oxygen.

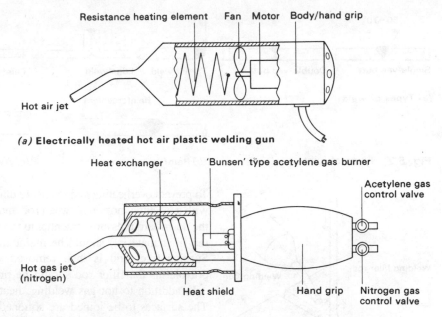

Resistance heating element   Fan   Motor   Body/hand grip

Hot air jet

(a) **Electrically heated hot air plastic welding gun**

Heat exchanger   'Bunsen' type acetylene gas burner

Acetylene gas control valve

Hot gas jet (nitrogen)

Heat shield   Hand grip   Nitrogen gas control valve

Fig. 5.1 Plastic welding guns   *(b)* **Gas heated hot nitrogen plastic welding gun**

Oxidation of the joint weakens it. Either type of gun can use air or nitrogen. The electrically heated gun is usually used in a workshop, whereas the gas heated gun is used on site.

Unlike metals which have a fairly sharply defined melting point, thermoplastics usually have a wide range of temperatures between which they start to soften and eventually degrade. The easiest plastics to weld are polyvinyl chloride (PVC) and polyethylene (PE) as they have a wide softening range. The basic technique is to apply a jet of heated air or nitrogen gas into the joint so that the edges of the parent plastic sheet are softened. Filler material, in the form of a rod of the same material as that being welded, is added into the joint in much the same way as when welding metals. Some degradation inevitably occurs, so that the strength of the joint is slightly below that of the surrounding material.

The preparation of the joint edges for welding thermoplastics is similar to that for metals. Figure 5.2 shows some typical joints. Single and double-vee joints have an angle of between 50° and 70°. A small root gap is provided. A feather edge should not be left as it would overheat due to the low thermal conductivity of the plastic, thus a nose of approximately 1 millimetre should be provided as shown in Fig. 5.2(*c*). Unlike metal welding, the bead or 'reinforcement' standing above the surface of the sheet should not be removed as it can increase the joint strength by as much as 20% (see Fig. 5.2(*b*)).

Unlike metal welding, the plastic filler rod is softened but not melted. It is presented to the work at right angles as shown in Fig. 5.3, the welding jet being kept between 10 and 20 millimetres from the joint.

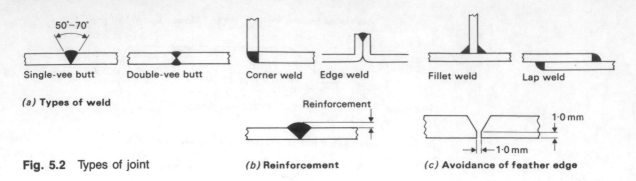

(a) **Types of weld**

Single-vee butt   Double-vee butt   Corner weld   Edge weld   Fillet weld   Lap weld

Fig. 5.2 Types of joint   *(b)* **Reinforcement**   *(c)* **Avoidance of feather edge**

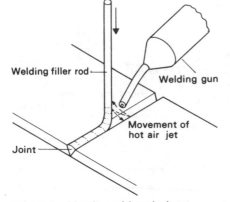

Fig. 5.3 Plastic weld technique

To prevent overheating and to ensure uniform softening the jet is moved with a weaving motion between rod and sheet. At the end of the weld, the downward pressure (essential to the merging of the joint edges with the filler material) should be maintained until the joint has cooled. Surplus filler rod is then removed with a knife. When welding polyolefins the filler rod should be presented to the work at 45°.

In addition to hot gas welding, heated tool welding is also used. The surfaces to be joined are softened against a heated surface and immediately brought together under pressure to make them bond. The surface of the heat source should be coated with a non-stick material such as 'teflon'. When welding thick sheet (6 mm or thicker) by this technique great care must be taken to avoid trapping air in the joint which would create discontinuities in the weld and severely weaken the joint. Thus, for thick material, it is preferable to 'shear' the edges together after heating to ensure perfect contact throughout the length of the joint.

5.3   Friction welding

When two surfaces are rubbed together without a lubricant the friction between them results in a conversion of mechanical energy into heat energy at the interface. This is exploited in friction welding (also called 'spin welding') by rotating one component against a stationary component until the joint faces reach their welding temperature. At this point rotation ceases and the axial pressure is increased to cause welding to take place as shown in Fig. 5.4. Unfortunately the heating effect will not be uniform. Overheating tends to occur at the periphery of the material where the maximum relative velocity between the two surfaces exists, and lack of heating occurs at the centre where the relative movement is minimal. Thus this technique is best applied to tubular rather than solid rod components and is widely used for joining plastic piping. Figure 5.5 shows some typical friction joint forms. It can be seen that the centre of the joint faces is either removed (Fig. 5.5(*a*)) so that the strength of the joint can be calculated more easily and accurately, or the centre of the joint faces form a mechanical joint or key (Figs. 5.5(*b*) and 5.5(*c*)). Both the pressure and the rotational speed must be carefully controlled and adjusted to suit the plastic being welded if satisfactory and consistent results are to be achieved.

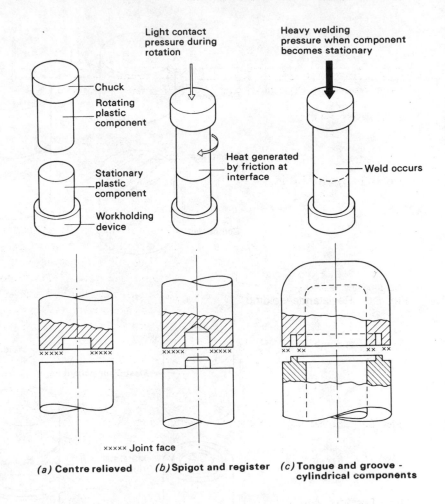

Fig. 5.4 Friction welding — principle

Fig. 5.5 Types of friction weld

××××× Joint face

(a) **Centre relieved**   *(b)* **Spigot and register**   *(c)* **Tongue and groove - cylindrical components**

5.4 Hot wire (resistance) welding

The problems of butt welding solid rod by friction heating have already been discussed. Further, external heating is equally ineffectual since all plastic materials are poor conductors of heat. A solution to the problem is shown in Fig. 5.6. An electrical resistance 'mat' of nichrome wire is laid in the joint and connected to an external power source. As soon as the joint faces have softened the supply is disconnected and pressure is applied. When the plastic has cooled and the weld is complete, the leads are cut off flush with the joint and the resistance 'mat' is left embedded in the joint.

5.5 Induction welding

Since plastic materials are insulators it may seem strange to consider the possibility of using electric induction heating to form a weld. The solution is quite simple. A ring of metal foil is laid in the joint and an induction coil is placed round the joint as shown in Fig. 5.7. A high frequency alternating current is passed through the coil and this induces a current in the foil ring embedded in the joint. The ring heats

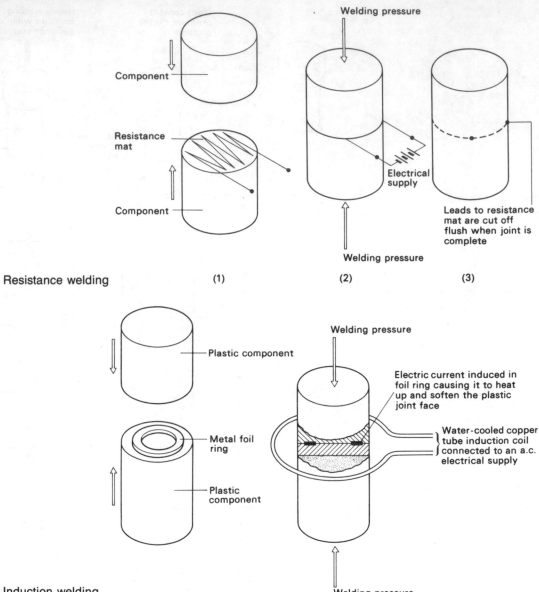

Fig. 5.6 Resistance welding   (1)   (2)   (3)

Fig. 5.7 Induction welding

up and softens the joint faces. Pressure is applied and the weld occurs, leaving the foil ring embedded in the joint. Obviously components with this type of joint must never be used in the presence of an alternating current field as the ring would heat up again and the joint would fail.

5.6 Di-electric welding

This process exploits the insulating properties of plastic materials. In this instance the plastic surfaces to be joined form the di-electric of a capacitor placed between two electrodes as shown in Fig. 5.8. A high

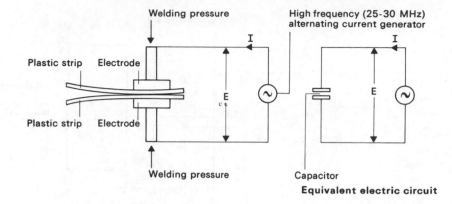

Fig. 5.8 Di-electric welding

Equivalent electric circuit

frequency alternating current of about 30×10^6 Hz is applied to the electrodes and the plastic commences to heat up from the inside. When the required temperature is reached, the current is switched off and the electrodes apply the pressure to complete the weld. The advantage of this technique is that the heating effect occurs uniformly within the plastic materials themselves and overcomes the problem of externally heating materials which have poor thermal conductivity. Not all plastic materials have the electrical properties to lend themselves to this process. The following plastic materials *cannot* be welded by dielectric techniques:

> Polyethylene (PE);
> Polypropylene (PP);
> Polycarbonate (PC);
> Polytetrafluoroethylene (Teflon) (PTFE);
> Polystyrene (PS).

5.7 Ultrasonic welding

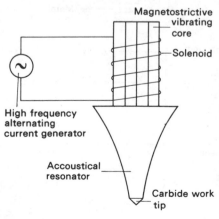

Fig. 5.9 Ultrasonic tool

This technique uses very high frequency sound waves (20×10^3 Hz or higher) to heat the joint interface. The term 'ultrasonic' means sound waves whose frequency is above the range of normal hearing. Ultrasonic techniques induce heat in the joint by friction as the surfaces vibrate rapidly together. Figure 5.9 shows the principal components of an ultrasonic welding tool. The 'horn' or accoustical resonator is designed to match the characteristics of the magnetostrictive core to the work head (tip), that is, the 'horn' acts as an accoustical tuning device. Magnetostriction is the property of the core material to expand and contract in sympathy with the alternating electromagnetic field produced by the solenoid and generator. The work head, or tip, is contoured to match the joint being made and is usually of a wear- and heat-resistant material such as tungsten carbide or titanium. Its mass must be kept very small so that it does not damp the high frequency vibration.

Most thermoplastics can be welded by ultrasonic techniques, the exceptions being the vinyls and the cellulosics. Figure 5.10 shows some

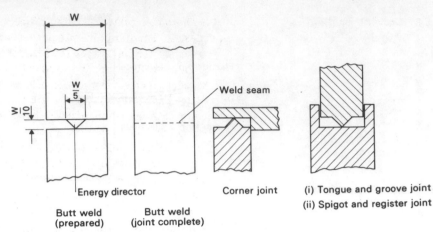

Fig. 5.10 Weld preparation

Energy director

Butt weld
(prepared)

Butt weld
(joint complete)

Corner joint

(i) Tongue and groove joint
(ii) Spigot and register joint

typical joints and it can be seen that they are formed with *energy directors* to localise the heat generated. This restricts the energy used and the volume of plastic softened. The small amount of softened plastic spreads out into the joint under pressure and produces the bond. The joint should be held for between a half and one second after current has been switched off for the plastic to solidify. In addition to welding, ultrasonics can also be used for joining plastics to non-plastic materials by the process of staking as shown in Fig. 5.11(*a*), and also for implanting metal inserts into plastic components as shown in Fig. 5.11(*b*).

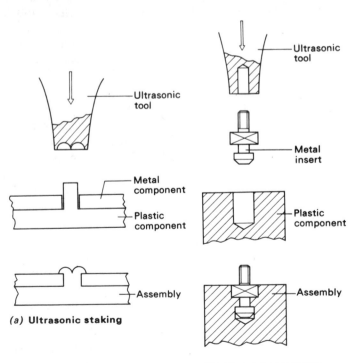

(a) Ultrasonic staking

(b) Metal insert fixing

Fig. 5.11 Further applications of ultrasonics

5.8 Solvent welding

As with thermal welding, this process can only be applied to thermoplastics. Thermosetting plastics *cannot* be solvent welded. Instead of softening the joint faces by the application or generation of heat, a suitable solvent cement is used. The surfaces to be joined are pressed together after the application of the solvent until evaporation is complete. Thus a highly volatile solvent is required to ensure quick and complete evaporation, and care must be taken to ensure adequate fume extraction. Many solvents give off highly flammable and toxic fumes and great care must be taken in their storage and use.

Bodied cements: These are used where gaps occur in the joint faces, since a volatile cement cannot fill gaps when it evaporates. A typical bodied cement is made by dissolving cellulose nitrate in amyl acetate. All bodied cements are made by dissolving some of the parent material in the solvent before applying the solvent to the joint faces.

Monomeric cements: These are made from the same monomers as the materials being joined. For example methylmethacrylate may be used to join any proprietary PMMA material. A catalyst is added to the solvent immediately before application and the bond is produced by polymerisation rather than by evaporation.

Joints made with solvent cements are weaker than those produced by thermal welding because of the shrinkage stresses set up by evaporation of the solvent. Furthermore, evaporation of the solvent may also create voids in the joint. Another problem is the difficulty in ensuring complete evaporation when large surface areas are being joined.

5.9 Adhesive bonding

The principles involved in the use of adhesives to bond materials together has already been discussed in Chapter 4. When used to join plastic components care is required to match the properties and composition of the adhesive to those of the materials being joined. Also the surface texture of many plastic materials prevents an adequate key being formed between the adhesive and the parent material. Furthermore, the adhesive must match the rigidity of the components being joined.

Elastomeric adhesives are used to join flexible materials.
Thermosetting adhesives are used to join rigid materials. These adhesives are generally stronger than elastomeric adhesives.

Elastomeric adhesives are usually based on natural, synthetic, or reclaimed rubber, which is dissolved in a solvent. Neoprene rubber is used for impact adhesives which form a bond on contact.

Thermosetting adhesives are insoluble in most solvents and are usually set by polymerisation after the addition of a suitable hardener (catalyst). Two methods of application are employed.

1. Resin and catalyst are mixed prior to use and curing starts at once. Thus the mixture must be applied quickly before polymerisation advances too far and a poor joint occurs. This technique is employed for moulding glass fibre reinforced components.

2. The resin is applied to one of the joint surfaces and the catalyst is applied to the other joint surface. Curing does not occur until contact is made between the two surfaces. This enables work to proceed at a more leisurely pace and is useful when manipulating large components and panels into position.

Epoxy resins are the strongest (and most expensive) of the thermosetting adhesives. They have incomparable strength and weathering characteristics and can be used to bond ill-fitting joints as very little shrinkage occurs, thus combining the properties of adhesive and void filler. Epoxy resins will be considered further in Section 5.15.

5.10 Heat bending techniques

Simple, straight bends in thermoplastic sheet follow the same principles as for sheetmetal working, the only difference being that the plastic material has to be heated before bending and that bending must take place whilst the material is still hot. It must then be held in position until it cools. For this reason the bending tools must be faced with materials having a low thermal conductivity such as wood or tufnol.

Strip heaters may be used to ensure that heating is localised along the line of the bend. This makes the plastic sheet easier to handle. Rapid cooling is required immediately after the bend is complete to avoid loss of shape and degradation of the plastic. Thick sheet should be heated on both sides prior to bending because of the low thermal conductivity of plastic materials. Like metals, some spring-back occurs when the plastic is removed from the bending tools and some degree of *overbend* will be required to counter this. The degree of overbend has to be determined by trial and error as it depends upon the properties of the plastic, its thickness, temperature, radius of bend, and rate of cooling. Other forming techniques to produce more complex shapes are outlined in Fig. 5.12. In all these examples the thermoplastic sheet is softened by pre-heating before forming.

Fig. 5.12 Forming plastic sheet material

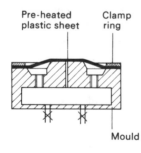

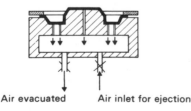

(a) Vacuum forming

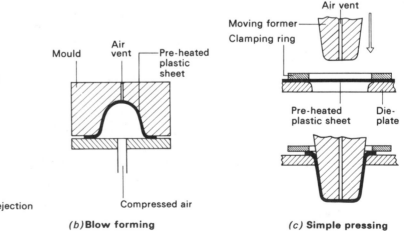

(b) Blow forming

(c) **Simple pressing**

5.11 Bench and machining operations for plastics

Although plastic components are usually moulded to their finished shape, it is sometimes necessary to resort to fitting and machining processes. For example:

(a) drilling holes from the solid where coring is inconvenient or the quantity does not warrant the use of complex moulds;

(b) sizing cored holes where shrinkage after moulding is unacceptable;

(c) tapping holes to produce internal screw threads;

(d) turning bushes, screws and other small components from solid rod where moulding is inappropriate;

(e) drilling holes in printed circuit boards so that the connecting wire of the components can be inserted prior to soldering;

(f) the machining, by normal workshop methods, of engineering components from Tufnol (reinforced plastic bars, rods, tubes, sheet, etc.);

(g) the *in situ* erection of ducting and equipment cabinets made from sheet plastic.

The hand tool operations usually associated with metalwork also apply to most plastic materials, except that the cutting forces involved are very much lower. Sheet plastic and extruded sections can be cut with a hacksaw, and thin thermoplastic materials can be cut with a strong craft knife. Thermoplastic materials can also be cut using a 'hot-wire' technique. An electrically heated wire coated with a non-stick material can be used to melt its way through the plastic material. Plastics can also be filed to shape, although this is more successful with thermosetting plastics and reinforced plastics as they are generally more rigid. However, rigid PVC and 'Perspex' can also be filed successfully. The rigid plastic materials can also be threaded using normal taps and dies. For the softer materials self-tapping screws are normally used.

The problems associated with machining plastics are as varied as the materials themselves. Within a broad framework of recommendations, each material needs to be treated individually and the optimum maching conditions have to be determined by experience. Just as plastic materials can be divided into two main groups of thermoplastics and thermosetting plastics, so the machining characteristics divide into two main groups.

Thermoplastics

Heat generated during cutting tends to build up due to the poor thermal conductivity of all plastic materials. This softens the material being cut so that it may collapse and weld itself to the tools. For the same reason the swarf produced may clog the tools — particularly the flutes of twist drills. Cooling by air blast is recommended: a liquid coolant cannot be used. The cutting speed may also have to be reduced to avoid overheating and softening despite the machine being worked below its optimum capacity.

Thermoplastics have low moduli of elasticity and high elastic recovery and will, therefore, deform under the cutting forces of the tool and spring back when the tool has passed. Thus the cutting forces must be kept low, or the cut must be repeatedly cleared (see 'woodpecker feed', Section 5.12).

From the above machining characteristics, it is obvious that tooling with very free-cutting characteristics is required. Particular attention should be paid to the flow of the chips produced, as thermoplastics are ductile and produce a long, continuous, ribbon-like chip.

As thermoplastic materials are soft, ductile and weaker than most metals, the cutting forces on the tool are lower and an acute rake angle can be used to knife-off the swarf freely. This prohibits the use of carbide and ceramic tooling. Super-high speed steel is suitable for most applications, but stellite is to be preferred for long production runs.

Cutting tools should be lapped to a high surface finish after grinding to reduce friction and the consequent possibility of the swarf adhering to the rake face.

Thermosetting plastics

Poor thermal conductivity again results in a tendency to overheating, but with thermosets softening does not occur. Thermosetting plastics tend to be rigid and brittle compared with the thermoplastics previously described. Thus they machine more like cast iron, producing a granular, discontinuous chip. This requires vacuum chip disposal facilities. The filler materials used with thermosetting plastics result in high abrasion and wear of high speed steel cutting tools. Fillers such as glass fibre, paper, mica and clay are particularly abrasive.

From the above characteristics it is apparent that low rake can be used as a discontinuous chip is produced. At the same time, the material requires a very hard, abrasion resistant cutting tool. Thus carbide or ceramic tooling is preferable when machining thermosets. Since low rake tooling increases the temperature at the cutting zone, the cutting speed has to be kept relatively low (30 to 50 m/min) to prevent thermal degradation of the plastic.

5.12 Drilling plastic materials (thermosetting)

As has already been indicated, the problems encountered when drilling plastic materials are completely different from those encountered when drilling metals. The situation is further complicated by the fact that the problems encountered when drilling thermosetting plastics are different to those encountered when drilling thermoplastic materials. Again, within these main groups are hundreds of individual material specifications each requiring special consideration. Thus the following notes are a generalisation of the main problems and their solution within broad material groupings. For thermosetting plastics these can be summarised as follows.

Heating effects

When drilling metal components much of the heat generated when cutting is dissipated by conduction through the body of the component. (Metals are relatively good conductors of heat.) Plastic materials are heat insulators, therefore the heat generated by cutting is trapped in the hole and can only be conducted away through the drill. This results in the drill overheating and the component burning. Liquid coolants cannot be used, but an air blast is useful in cooling the drill and the component. It also blows the swarf towards the suction extractor as shown in Fig. 5.13.

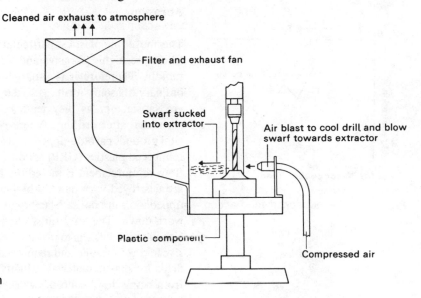

Fig. 5.13 Air blast and extraction

Shrinkage

However much care is taken, heat is generated in the hole whilst drilling. This causes the component to expand and shrink again on cooling. There is also appreciable deformation of the material at the point of cutting and spring back when the cutting edge has passed. The combination of these effects can be so pronounced that oversize drills frequently have to be used to produce a hole of the required size. For example:

2 BA tapping size drill — steel   3.7 mm diameter
2 BA tapping size drill — thermosetting plastic 3.9 mm diameter

Feed

Two problems affect the rate of feed. First, the heating effect already mentioned, and second the tendency for the swarf to clog in the flutes of the drill. To overcome these difficulties the drill is fed into the work

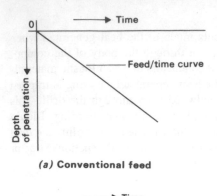

(a) Conventional feed

(b) 'Woodpecker' feed

Fig. 5.14 'Woodpecker' feed action

with a 'woodpecker' action, that is, instead of a continuous uniform rate of feed as usually employed when drilling metals, the drill is fed into the work quite rapidly for a short distance and then withdrawn momentarily so that the plastic can 'breathe'. This allows the air blast to cool the drill and eject the swarf. It also allows the hole to shrink in size. This shrinkage is cleared as the drill is given its next increment of feed, and prevents the hole closing on the body of the drill and so prevents the drill and the workpiece overheating. This method of feeding the drill is shown diagrammatically in Fig. 5.14.

Abrasion

Thermosetting plastics, particularly those with mineral or fibre-glass fillers are highly abrasive and wear out high speed steel drills very rapidly. Therefore it is usual to use carbide-tipped drills. Usually ordinary 'masonry' drills as sold by ironmongers for drilling brickwork are satisfactory as they tend to drill oversize and this allows for shrinkage after drilling. Where precision holes are required, precision ground, carbide-tipped drills are used. When drilling glass fibre reinforced plastics (GRP) solid carbide drills must be used. The glass fibre reinforcement is incredibly abrasive and high-speed steel drills are destroyed very quickly by being worn undersize. Even carbide-tipped tools are not satisfactory as the steel body of the drill is rapidly worn down. The only satisfactory way to drill GRP components on a production basis is to use solid carbide tools. These are very expensive and very fragile and require careful and expert handling. Whereas drills for plastic materials usually have a slow helix to prevent them from being drawn into the workpiece, drills for GRP have a 'quick' helix to help eject the swarf rapidly. The material offers sufficient resistance to cutting to prevent the drill being drawn into the work, and the quick helix removes the swarf freely to reduce wear on the drill body.

Cutting speed

This tends to be kept fairly low to reduce the heating effect and wear of the flutes from abrasion. For holes up to 12 millimetre diameter, the cutting speed can be in the range 35 m/min to 45 m/min depending upon the type of plastic being drilled and the filler material used.

Work-holding

Components made from thermosetting plastic materials tend to be rigid and brittle, so great care must be taken in clamping to avoid cracking them by using too great a clamping force or distorting the component.

5.13 Drilling plastic materials (thermoplastic)

Since there is a very wide range of thermoplastic materials varying equally widely in properties, it is only possible to give a general introduction to the main problems encountered when drilling this group of materials and to suggest possible solutions.

Heating effects

As previously stated, all plastic materials are good thermal insulators and thermoplastics are no exception. Therefore the heat generated during cutting cannot be conducted away through the body of the component. This causes the material to overheat rapidly, soften and weld itself to the drill. This prevents the drill cutting properly so overheating and welding builds up until the drill is dragged to a halt or is broken, both the drill and the component being scrapped. Liquid coolants and lubricants cannot be used, only an air blast is suitable.

Shrinkage

This effect is even more acute with thermoplastics than in the case of thermosetting plastics and this must be taken into account when selecting drill sizes where the finished diameter of the hole is critical. There is also the problem of the material closing on the body of the drill. This adds to the heating problems outlined above and the additional friction increases the overheating and results in total clogging of the drill flutes, breakage of the drill, and scrapping of the work.

Feed

Both of the problems outlined above can be overcome by the use of drills specially desgined for these materials and by the use of 'woodpecker' feed associated with the correct cutting speed. The importance of 'woodpecker' feed has already been discussed in Section 5.12. This action needs to be even more exaggerated when drilling thermoplastics. The drill is run much faster (there is no abrasive filler) and each feed increment is very coarse, resulting in a 'bodging' action. This results in the required amount of material being removed before the temperature can build up to the softening point for the plastic concerned. Between each increment of feed the drill point can cool down in the air blast and the component can 'breathe' whilst cooling down and shrinking back. The next feed increment clears the shrinkage (spring-back) as the drill re-enters the hole and binding on the body of the drill becomes non-existent. It is usual to employ special high-speed steel tools which have been marketed for drilling thermoplastic materials. These drills have slow helix angles, very wide flutes, thin webs, and acute points (80°). The flutes are polished to reduce friction and clogging. They are ground with a generous clearance angle and are given a keen cutting edge. The total geometry of the drill is designed to enable it to cut as freely as possible and to allow the swarf to have rapid egress from the hole, so as to keep the heat generated to a minimum.

Abrasion

With very few exceptions thermoplastic materials do not contain abrasive fillers, therefore high-speed drills perform satisfactorily provided they have the correct geometry. It is also possible to give high-speed steel tools a narrower wedge angle and a keener edge than carbide tooling. Not only is HSS tooling more economical in first cost, but it is easier to service high-speed steel tools and maintain the keen cutting edge necessary to ensure free cutting without overheating.

Cutting speed

Since most thermoplastic materials are not abrasive, they can be cut with a much higher speed than for thermosetting plastics. For holes up to 12 millimetres diameter the cutting speed could be in the range of 200 to 250 m/min for carbide drills and 80 to 100 m/min for high-speed steel drills. An acute 'woodpecker' feed action prevents the heat build up normally associated with high-speed cutting. This combination of high cutting speed and coarse 'woodpecker' feed is geared to rapid material removal and economical production without the problems associated with overheating and softening of the workpiece material.

Work-holding

Components made from thermoplastic materials are less rigid and brittle than those made from thermosetting plastics. Therefore they are less likely to be cracked when clamped into work-holding fixtures. However they do tend to be easily distorted by their inherent flexibility, and this can lead to inaccuracy if care is not exercised when clamping. Where only small holes are being drilled, it is quite normal to 'hand-hold' the component in a simple fixture as the cutting forces are sufficiently low as not to endanger the operator.

5.14 Turning plastic materials

Only the more rigid materials are turned in the lathe, for example 'Tufnol' rods and sections, and extruded rods of thermoplastic materials such as polypropolene, nylon, 'Perspex', etc. The 'Tufnol' range of materials can be turned using the same tools, cutting angles, and speed and feeds, as for mild steel. However being abrasive, carbide tooling gives a longer life between regrinds. Positive rake tooling should be used, even with carbide tooling. This is no problem since the cutting forces are relatively low compared with mild steel. A liquid coolant cannot be used. Vacuum extraction is recommended as the dust produced can cause irritation and sensitisation. Alternatively a face-mask should be worn.

When turning thermoplastics the same problems exist as when drilling. High rake angles are used to 'knife-off' the swarf and the rake face should be polished to avoid chip welding. The cutting speed is kept down not because of tool wear, but to avoid overheating and softening the plastic itself. The feed should be interrupted to break up the swarf, and a number of light cuts at fairly fast feed rates should be used to avoid overheating and to allow the plastic to 'breathe' and swell due to 'spring-back' after each cut. Free-cutting is essential with these materials.

5.15 Casting plastic materials

Metals are cast by raising them above their fusion temperatures, pouring them into moulds and waiting for them to freeze into solid components. The casting of plastics is entirely different. Basically, a low viscosity resin is mixed with a catalyst and poured into a mould that has been treated with a suitable release agent. After curing has taken place, the mould is opened and the casting is removed. Three groups of plastic materials (epoxides, phenolics, polyesters) are used for casting.

(a) *Epoxides*, where maximum strength is required for tooling or where maximum electrical insulation is required for insulators and encapsulating electrical components.

(b) *Phenolics*, where appearance is important and the inherent brittleness is relatively unimportant (e.g. billiard and snooker balls).

(c) *Polyesters*, for hobby casting as well as for a range of industrial and construction industry products, such as synthetic marble, stone and brick veneers used for interior trim, shop and bar fitting. The high viscosity, high exotherm polyesters developed for glass fibre moulding is not suitable for casting. Special low viscosity, low exotherm polyesters have been formulated for cast applications.

Since this chapter is concerned mainly with the engineering applications of plastic materials, cast epoxides will now be considered in rather greater depth. Epoxy resins have oxygen and aromatic rings in the polymer chain and this results in the superior properties of these resins. They are available in a wide range of formulations given an equally wide range of properties and curing characteristics, leading to an equally wide range of applications. Since no water or volatile substances are produced during the curing cycle, little shrinkage occurs and epoxide castings have high dimensional stability. They also make excellent gap filling adhesives and have been considered as such in Section 4.9. Two additional applications are considered here, the casting of *tooling epoxies* and the *encapsulation* of electrical and elec-tronic components.

(*a*) *Tooling epoxies* are usually reinforced with aluminium powder, and are cast to make large and complex forming dies used for short run sheet metal pressings. Aluminium-filled epoxy resins are easier to machine and lighter to handle than tool steels. Neither is costly heat treatment of large dies necessary. Tooling epoxies are also readily repaired and can be easily altered by building up additional layers of resin.

Other applications are casting patterns and models for three dimen-sional copy milling where their light weight and impact resistance facilitates handling and their corrosion resistance facilitates storage. Lead time between initial design and production is greatly reduced by using tooling epoxies.

(*b*) *Encapsulating electrical and electronic components* is now widely used especially for equipment designated for tropical and other electrically hostile environments. Epoxy resins are rapidly replac-ing traditional insulating materials because of their improved elec-trical and mechanical properties. For example, epoxy resins are replacing Chatterton's compound for 'potting' or encapsulating transformers and chokes. They are also replacing ceramic (porcelain) insulators for heavy duty power transmission equip-ment. The epoxy insulators having greater impact strength.

For the impregnation of transformers, or motor and generator stators and rotors, unfilled epoxides are used, specially formulated to have very low viscosities. They are solvent free and have good wetting pro-perties so that voids are unlikely to occur and the electrical function is improved. Two-part systems that cure, on mixing, at ambient temperatures are used for casting and encapsulation. For impregna-tion the epoxides are formulated to withstand a pot temperature of 20 °C for up to fourteen days. The components are dipped at 20 °C which causes full penetration and drives off any entrapped moisture. Curing takes upwards of two hours at 120 °C and is achieved by stov-ing the dipped components in heated ovens.

When making a casting or encapsulating electrical components, the resin must be poured into the mould so as not to entrap air and create voids. The thermoset should be poured slowly and steadily down one side of the mould (which must be adequately vented). The mould should be vibrated to assist air bubbles to rise to the surface where they can be pricked and burst.

Problems

Section A

1 Thermoplastic materials can be welded using:
(*a*) an oxy-acetylene torch;
(*b*) arc welding equipment;

(c) a hot air gun;

(d) a brazing torch.

2 When drilling plastic materials the feed of the drill should be:

(a) slow but steady;

(b) fast but steady;

(c) fast with a woodpecker action;

(d) very slow with a woodpecker action.

3 A suitable coolant when machining plastics is:

(a) compressed air;

(b) suds;

(c) water;

(d) paraffin.

4 Carbide-tipped tools are used when machining thermosetting plastics because:

(a) of the high temperature involved;

(b) the fillers used are abrasive;

(c) they chip less readily;

(d) they are easy to regrind.

5 To ensure that sheet plastic takes a permanent set when bent to shape it should be:

(a) scored with a knife along the bend line;

(b) painted with a solvent prior to bending;

(c) cooled prior to bending;

(d) heated prior to bending.

Section B

6 With the aid of sketches describe how four of the following joining techniques are applied to plastic materials:

(a) hot gas welding;

(b) friction welding;

(c) hot wire (resistance) welding;

(d) induction welding;

(e) di-electric welding;

(f) ultrasonic welding.

7 (a) Epoxides are an important industrial group of plastic materials. Describe in detail two applications of cast epoxides.

(b) Epoxides also form the basis of high strength adhesives. Describe how a joint should be prepared and completed using epoxide adhesives.

8 (a) Explain how the machining of thermoplastic materials varies from the machining of metals.

(b) Explain what precautions must be taken when machining thermosetting plastics to ensure a reasonable tool life.

9 With the aid of sketches, describe how thermoplastic materials may be formed by:

(a) vacuum forming;

(b) blow forming;   (c) pressing.

10 Explain the meaning of the following terms as applied to the solvent welding and adhesive bonding of plastic materials:

(*a*) bodied cement;

(*b*) monomeric cement;

(*c*) elastomeric cement;

(*d*) impact adhesive;

(*e*) thermosetting cement.

6 Electrical principles and practice

6.1 Basic principles

Figure 6.1 shows a simple electrical circuit which contains all the fundamental requirements:

(a) a source of supply — in this case a battery;
(b) a means of controlling the flow of electricity — in this case a switch;
(c) a means of overcurrent protection — in this case a fuse;
(d) a load (appliance) which converts the electrical energy into useful work — in this case a filament lamp;
(e) suitable conductors (wires) to connect the above items together to form a circuit.

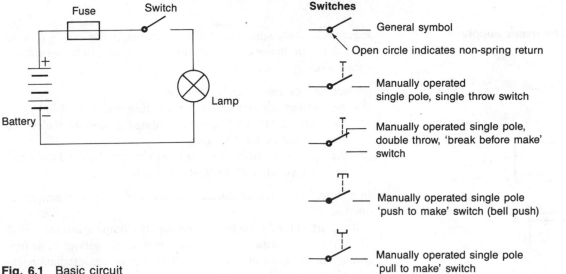

Fig. 6.1 Basic circuit

Figure 6.2 shows a number of filament lamps *wired in parallel* with each lamp or group of lamps controlled by its own switch. This type of circuit is used for wiring the lighting circuits in a house and the source of supply is the alternating current mains. The parallel circuit is used because it has the advantage that if one lamp fails, the remaining lamps stay alight. In this example the lighting circuit is connected to the distribution fuse board (DFB) of a house or office block to form a *final circuit*. Note that 'final circuits' are sometimes referred to as 'subcircuits'. *A final circuit can be defined as: 'an assembly of conductors and accessories (switches, lamps, etc.) emanating from a final distribution fuse board'.*

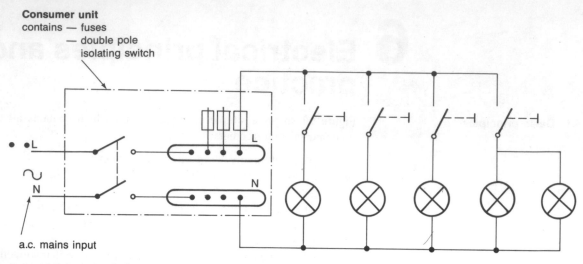

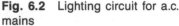

Fig. 6.2 Lighting circuit for a.c. mains

6.2 The mains supply

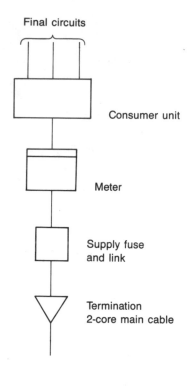

Fig. 6.3 Supply intake

Figure 6.3 shows, schematically, the incoming mains supply (2-wire) for a domestic house or a small office block. It can be seen that it consists of:

(*a*) the service cable, terminating in;
(*b*) the 'cut-out' containing the service fuse and neutral link;
(*c*) the 'meter' for measuring and recording the quantity of electrical energy used by the consumer;
(*d*) the 'consumer unit' which combines the main linked double-pole isolating switch with the distribution fuse board.

In a larger, industrial or commercial installation these functions are often separated.

Items (*a*), (*b*) and (*c*) belong to the Supply Company and are sealed. The service fuse in the cut-out not only protects the service cable from damage in the event of a major fault arising in the installation, but allows the Supply Company to disconnect the supply from the premises in the event, say, of the consumer's account not being paid. Item (*d*) is provided by the consumer and should be situated as close as possible to the service cable and meter so that the main cables to the consumer unit (the 'tails') can be kept as short as possible.

For a final circuit, such as that shown in Fig. 6.2, to operate safely from a mains supply, the lamps, switches, wiring insulation, etc., must be suitable for operating at a potential of 240 volts. This is the nominal potential for a single-phase alternating current supply in the UK. This nominal potential may legally fluctuate by 6%. The nominal potential of 240 volts is the root-mean-square (r.m.s.) value as indicated by a voltmeter. The peak value of the supply voltage is very much higher as shown in Fig. 6.4.

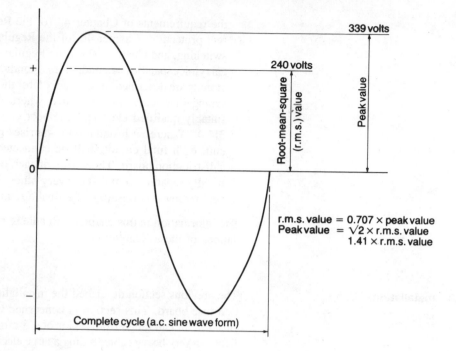

r.m.s. value = 0.707 × peak value
Peak value = √2 × r.m.s. value
          1.41 × r.m.s. value

Complete cycle (a.c. sine wave form)

Fig. 6.4 Peak voltage (a.c.)

$$\text{Peak value} = 1.414 \times \text{r.m.s. value}$$
$$= 1.414 \times 240 \text{ volts}$$
$$= \underline{339 \text{ volts}.}$$

This is the actual potential which the insulation has to withstand and it is the potential which causes electrical shock should any person come into contact with a live conductor. It is very dangerous and a shock at mains potential can be lethal. Therefore it is essential, when working at mains potential, that the installation is strictly in accordance with the safety codes and recommendations of the *Regulations For Electrical Installations* published by the Institution of Electrical Engineers. Section 314 of these regulations states:

314-1 Every installation shall be divided into circuits as necessary to
  (i) avoid danger and minimise inconvenience in the event of a fault, and
 (ii) facilitate safe operation, inspection, testing, and maintenance.

314-2 Separate circuits shall be provided for parts of the installation which need to be separately controlled, in such a way that these circuits are not affected by failure of other circuits.

314-3 The number of final circuits required, and the number of points supplied by any final circuit, shall be such as to comply with the requirements of Chapter 43 (of the Regulations) for overcurrent protection, Chapter 46 (of the Regulations) for isolation and switching, and Chapter 52 (of the Regulations) as regards current-carrying capacities of conductors. Standard arrangements of final circuits are described in Appendix 5 (of the Regulations), but other arrangements are not precluded where they are specified by a suitably qualified electrical engineer.

314-4 Where an installation comprises more than one final circuit, each final circuit shall be connected to a separate 'way' in a distribution board. The wiring of each final circuit shall be electrically separate from that of every other final circuit, so as to prevent indirect energisation of a final circuit intended to be isolated.

The information in this chapter will adhere rigidly to the recommendations of these Regulations.

6.3 Installations

The previous section described the installation as far as the distribution fuse board. This section is concerned with the final circuits. A final circuit can range from a pair of 1.5 mm² cables feeding a single light to a very heavy cable feeding a large electric motor in an industrial installation. The one important rule which applies to final circuits is that mentioned in the Electricity Supply Regulation 27: 'All conductors and apparatus must be of sufficient size and power for the work they are called upon to do, and so constructed, installed and protected as to prevent danger.' The IEE Regulations ensure that this statutory regulation is complied with.

Chapter 43 in the IEE Regulations deals with the requirements for disconnecting circuits from the supply in the event of overload and short circuits (overcurrent protection). The rating of a protective device (fuse or circuit breaker: see Section 6.6) must not be less than the designed load current of the circuit and, also, that rating should not exceed the current carrying capacity of the lowest-rated conductor in the circuit. There are five important general groups of final circuits.

(a) Rating not exceeding 15 amperes

Table 4A in Appendix 4 of the IEE Regulations indicates some final circuits which have a rating not exceeding 15 amperes. The main requirements of such circuits is that the number of points which may be supplied is limited by their aggregate demand as determined by the data in Table 4A; there must be no allowance for diversity; and the current rating of the cable must not be exceeded. The interpretation of this regulation is shown in Fig. 6.5. *Diversity* is the term used to describe the fact that not all the loads connected to a circuit may be necessarily switched on at the same time. For instance it is highly unlikely that, in a private dwelling house, every light, every power

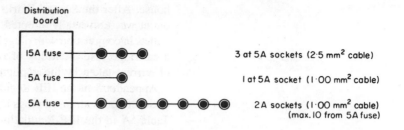

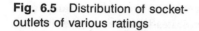

Fig. 6.5 Distribution of socket-outlets of various ratings

point, the cooker, immersion heater and other fixed appliances will be switched on at the same time. For this reason there is some relaxation of the rules for lighting circuits in private houses. From Table 4B in Appendix 4 of the IEE Regulations it can be seen that the *Diversity Factor* for the lighting circuit in a private house is 66%. Thus it is assumed that only two thirds of their connected load will be switched on at any one time, and the circuit is cabled and protected on this assumption. Domestic lighting circuits should be two in number. The reason for this recommendation is that if one circuit fuse 'blows', the other circuit will supply the remainder of the lights in the house, that is, allow sufficient lighting to enable repairs to be carred out. In practice, domestic lighting circuits are rated at 5 amperes. This means that they use 1 mm² or 1.5 mm² cable, are protected by a 5 ampere fuse or miniature circuit breaker (MCB), and that flexible cords from ceiling roses are sized at 0.5 mm² or 0.75 mm². In industrial installations the lighting circuits may be much more heavily rated and cabled and protected accordingly.

(b) Rating exceeding 15 amperes

Circuits with a rated capacity exceeding 15 amperes include those feeding cooker units and water heaters which are rated at 15/20 amperes. Final circuits over 15 amperes must not feed more than one point. Such circuits are, therefore, called *radial circuits* since each appliance is separately wired to its own individual fuse. There are two exceptions to this regulation. One is where there is a socket outlet integral with a cooker unit. The other applies to circuits feeding 13 ampere socket-outlets used with fused plugs.

(c) Rated for 13 ampere fused plugs

At one time all socket-outlets were of the round pin type rated at 15 amperes for power points and 5 amperes for reading lamps and other low current devices. Each socket-outlet was separately wired to its own fuse in the distribution board. That is, *radial circuits* were used. Such circuits were wasteful of cable and require very large distribution boards with many ways to provide all the services in a modern house. After the Second World War the standard 13 ampere socket-outlet was universally adopted. The 13 ampere circuits use socket-outlets in conjunction with plugs having rectangular pins and containing a cartridge fuse the rating of which can be matched to the amount of current taken by the appliance connected to the plug.

Appendix 5 of the IEE Regulations deals with circuits for socket-outlets fitted with fused plugs and rated at not more than 13 amperes. Table 5A in the IEE Regulations indicates how these socket-outlets are to be connected. It will be noted that a diversity factor has been applied. For instance, two 13 ampere socket-outlets can be wired by a single 2.5 mm^2 cable and protected by a 20 ampere fuse. This apparent contravention of the Regulations is permitted because it is considered unlikely that both 13 ampere socket-outlets would ever be called upon to deliver their full rated current at the same time. There is an even greater level of diversity allowed in a *ring-main* circuit where some socket-outlets never supply more than is needed for a television set, a radio, or a table lamp (i.e. less than 0.5 A).

Figure 6.6 shows a typical *ring circuit* which complies fully with the Regulations' requirements. Note that 13 ampere socket-outlets with fused plugs must only be used on an alternating current system. The symbols used in this figure are those used for house wiring diagrams instead of the, more familiar, theoretical circuit diagram. Note that

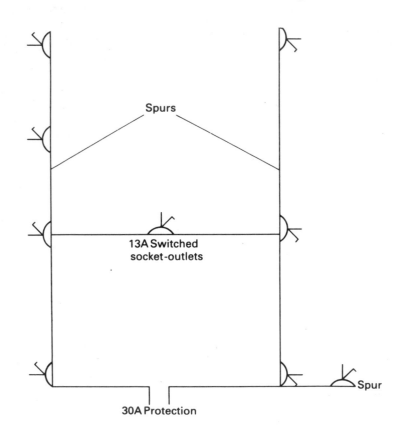

Fig. 6.6 13A ring main (wiring diagram) using 'architectural' symbols

each socket outlet is connected by a single line, which represents the two core and earth power cabling of the ring main. The maximum number of 13 ampere socket-outlets connected in one circuit will depend upon the estimated peak loading of the circuit. Table 5A of Appendix 5 of the IEE Regulations indicates that for domestic installations a single 30 ampere ring circuit may serve a floor area of up to 100 m^2 with an unlimited number of socket-outlets. Consideration should, however, be given to the loading in kitchens which may require a separate circuit. The total number of fused *spurs* which may be taken from the ring circuit is unlimited, but the number of non-fused spurs must not exceed the total number of socket-outlets and items of fixed (stationary) equipment connected permanently in the circuit. A non-fused spur feeds only one single or one twin socket-outlet or one permanently connected piece of equipment. Such a spur is connected to the ring circuit at the terminals of socket-outlets or at joint boxes or at the origin of the circuit in the distribution board. A fused spur is connected to the circuit through a *fused connection unit*, the rating of the fuse in the unit not exceeding that of the cable forming the spur and, in any event, not exceeding 13 amperes. The total permissible load of fixed appliances fed from a single final circuit must not exceed 13 amperes when protected by a fuse, or 16 amperes when protected by an MCB.

(d) Rated for feeding fluorescent-lamp circuits

One of the main requirements with regard to fluorescent lighting circuits is the switching. In such circuits, it is not sufficient to assume a maximum load of 100 watts per lamp-holder (as allowed for filament lamps), other factors must be considered. The inclusion of a choke (inductor) increases the current taken by the circuit. To comply with the IEE Regulations the nominal power rating of a fluorescent fitting must be multipled by not less than a factor of 1.8. This factor is based on an assumption that the light unit is corrected to a power factor of not less than 0.85 and takes into account control gear losses. IEE Regulation 537-19 states that any switch controlling circuits comprising discharge lamps should have a rating of not less than twice the steady current in the circuit it controls, unless the switch has been specially designed to break an *inductive load* at its full rated capacity. The recommended switch type is *quick-make/slow-break*, or else a.c. rated.

(e) Rated for feeding a motor

Final circuits which feed motors require some special consideration, though for the most part they must comply with the Regulations which apply to other types of final circuits. The current rating of cables feeding a motor should be based on the full-load current taken by the motor. More than one motor may be connected to a 15 ampere final circuit provided that the full-load rating of all the motors added together does not exceed the rating of the smallest cable in the circuit. If, however, the rating of the motor does exceed 15 amperes, then the circuit must supply one motor only. It must also be remembered that the starting-up current surge for a motor can often exceed the full-load current several times and this must be taken into account in the overcurrent protection system and the cabling used. IEE Regulations 552-1 to 552-4 apply particularly to motor circuits, the control of motors, and the overcurrent protection.

Figure 6.7 shows a typical arrangement of 'final circuits' which may emanate from the consumer unit for a domestic dwelling. It can be seen that both 'ring circuits' and 'radial circuits' are used. When selecting cables for any given installation, not only must they satisfy the requirements of the IEE Regulations for safe current carrying capacity, but they must also satisfy Regulation 522-8 which states that for domestic circuits the maximum permissible volt drop must not exceed 2.5%. The drop is measured between the supply terminals (at the supply intake position) and any, or every, point in the installation. That is, on full load, on a 240 volt supply, the maximum permissible volt drop is 6 volts.

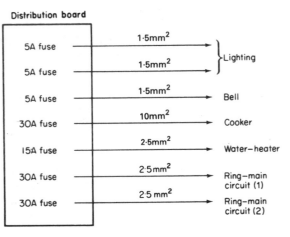

Fig. 6.7 Typical final circuits from a distribution board

6.4 Typical circuits

Figure 6.2 shows a number of lighting circuits wired in parallel. To wire up these circuits as drawn would, in practice, result in a large number of inconvenient connections. Figure 6.8 shows how the circuit can be wired in practice by *looping-in*. This may be done at the switch as shown in Fig. 6.8(*a*), or at the ceiling rose as shown in Fig. 6.8(*b*). Modern practice favours the latter system as it is more economical in cabling. However it requires a suitable ceiling rose containing three connector blocks and, since one of the connector blocks is permanently 'live', the connector blocks must be shrouded with suitable insulating material so that they cannot be accidentally touched

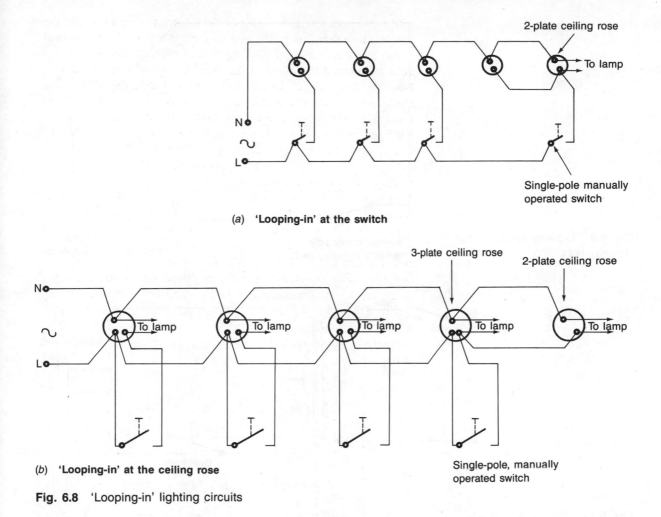

(a) **'Looping-in' at the switch**

(b) **'Looping-in' at the ceiling rose**

Fig. 6.8 'Looping-in' lighting circuits

when the ceiling rose is opened to allow replacement of the flexible cable.

Figure 6.9 shows how a light over a stairway may be wired so as to allow it to be operated by switches placed at the top and bottom of the stairway. This is called 'two-way' switching and requires single-pole/double-throw switches. Figure 6.10 shows a bell circuit powered by a mains transformer to reduce the potential to a suitable level (e.g. 12 volts). Figure 6.11 shows a 13 ampere ring main. Note the *earth continuity conductor*: see Section 6.8.

6.5 Terminations and connections

IEE Regulation 527-1 states:

'Every connection or joint shall be mechanically and electrically sound, be protected against mechanical damage and any vibration liable to occur, shall not impose any appreciable mechanical strain on the fixings of the connection, and shall not cause any harmful mechanical damage to the

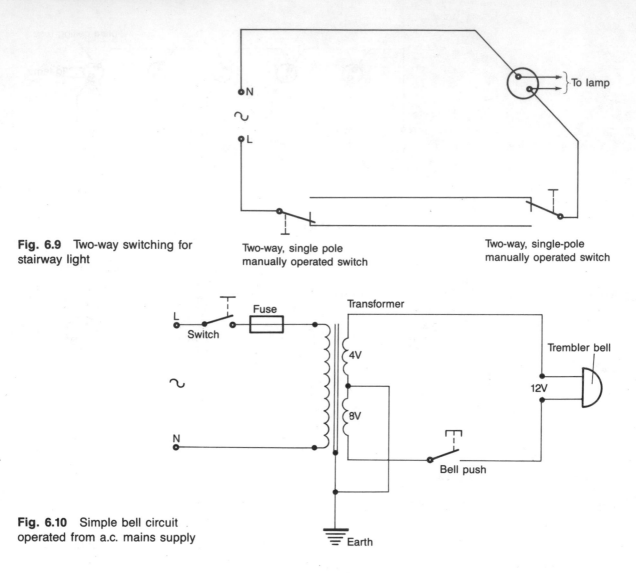

Fig. 6.9 Two-way switching for stairway light

Two-way, single pole manually operated switch

Two-way, single-pole manually operated switch

Fig. 6.10 Simple bell circuit operated from a.c. mains supply

cable conductor. Joints in non-flexible cables shall be made by soldering, brazing, welding, or mechanical clamps, or be of the compression type. All mechanical clamps and compression-type sockets shall securely retain all the wires of the conductor'.

The Regulation goes on to stipulate that terminations and joints shall be appropriate to the size and type of conductor with which they are to be used, and that terminations shall be suitably insulated for the voltage of the circuits in which they are situated.

There are a number of techniques for terminating conductors to comply with the above requirements and some typical examples will now be described.

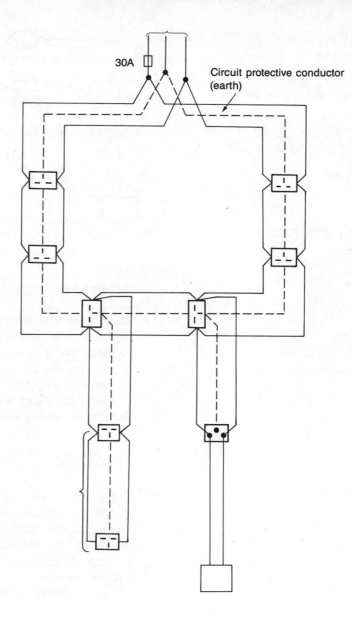

Fig. 6.11 13A ring main with earth conductor

Pinched screw

This method is used with most domestic electrical accessories. The insulation is stripped from the end of the cable for the required length, and the conductor is inserted into the terminal and secured by the grub screw(s) provided. Alternative light duty terminations are shown in Fig. 6.12(a).

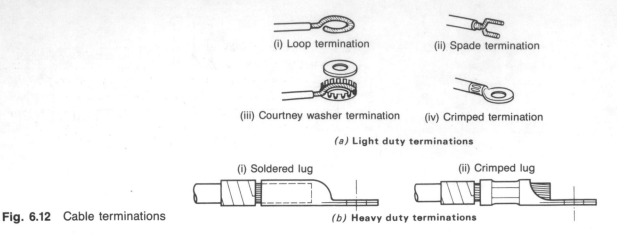

(i) Loop termination   (ii) Spade termination

(iii) Courtney washer termination   (iv) Crimped termination

(a) **Light duty terminations**

(i) Soldered lug   (ii) Crimped lug

Fig. 6.12 Cable terminations

(b) **Heavy duty terminations**

Soldered lug

These are used to terminate heavy duty cables before bolting them to bus-bars or the terminals of heavy duty switch gear or other devices. Figure 6.12(*b*(i)) shows a typical soldered lug. The actual technique of soldering was discussed in Chapter 4. The use of soldered lugs presumes the use of conductors made from materials such as copper which can be easily soldered. Aluminium is more difficult to solder and is more usually used with compression joints — see below.

Crimped lug

These are now tending to supersede soldered lugs as they can be equally applied to aluminium and copper cables. They are easy to fit on site and on preformed cables in the factory. There is no possibility of corrosion from the flux and no heat damage to the insulation. However care must be taken in their assembly onto the conductor to ensure a low resistance joint which will not work loose. Figure 6.12(*b*(ii)) shows a crimped (compression joint) lug.

Bolted connections

These are used to connect cables terminated in lugs to bus-bars and the terminals of heavy duty switch gear. They have the advantage of easy assembly and dismantling. However they do have some disadvantages: the bus-bar or the terminal has to be drilled to take the bolt and this reduces its cross-section and current handling capacity; contact pressure is also less uniformly distributed in bolted joints than in clamped joints and this tends to reduce the current handling capacity of the joint.

Mechanical clamping

A clamped joint is easy to make, no particular preparation being required other than removing the insulation and cleaning the conductors. The effective cross-sectional area of the conductor is not affected, but this type of joint does tend to be bulky. The joint is easily assembled and dismantled, but some care must be taken to ensure that the joint surfaces are in definite mechanical contact and that the bolts and nuts of the clamp are locked tight. Figure 6.13 shows bolted and clamped connections.

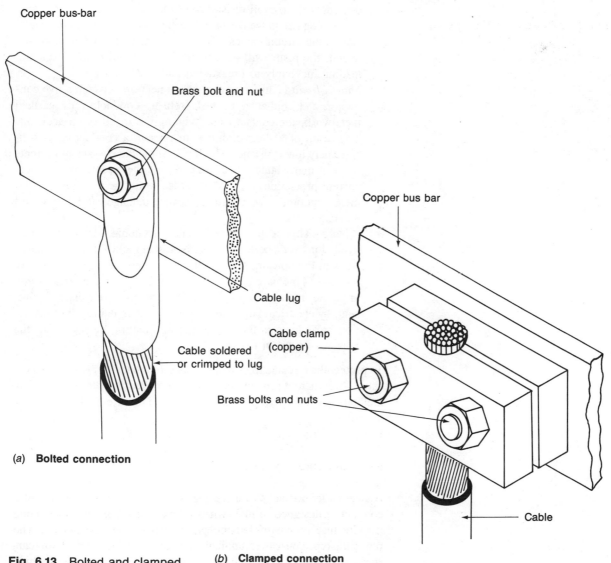

(a) **Bolted connection**

(b) **Clamped connection**

Fig. 6.13 Bolted and clamped connections

6.6 Electrical protection (overcurrent)

In electrical work, the term 'protection' is applied to the precautions taken to prevent damage to the various parts of an electrical circuit: wiring system, accessories, fittings, appliances and apparatus. The prevention of such damage, generally of a physical nature, means the prevention of danger to life, limb and property from shock and fire of an electrical origin. This is a very wide subject and, within the confines of this chapter, will be limited to *overcurrent protection* and *earthing* as applied to domestic premises and small offices. Overcurrent or excess current is the result of either an overload or a short circuit.

(a) *Overloading* occurs when an extra load is taken from the supply. This load, being in parallel with the existing load in a circuit, decreases the overall resistance of the circuit resulting in a corresponding rise in the current. This increased current immediately causes the circuit cables to heat up and, if the overload is sustained, the result will be a rapid deterioration in the insulation leading inevitably to breakdown and fire.

(b) *Short Circuit* is direct contact or connection between a live conductor and, either a neutral (return) conductor, or earthed metalwork, the contact usually being the result of an accident. The result of a short circuit is to present a conducting path of extremely low resistance which will allow the passage of currents often of many hundreds of amperes. If the circuit has no overcurrent protection, the circuit cables would heat up rapidly and melt, equipment would suffer severe damage, and fire would inevitably result.

(c) *Ageing*. This is the natural, and inevitable, deterioration of insulation over a period of time. Now that many electrical installations in the housing stock of the nation are getting old, the fire hazard has increased to a level where fires of an electrical origin are quite frequent. In fact about 20% of fires are caused, at present, by electrical faults, most of these being the result of insulation breakdown in the wiring through natural ageing or for the reasons set out in (a) and (b) above.

To protect the circuit cables from overcurrent hazards, from whatever cause, overcurrent protection must be provided in every circuit. The methods of overcurrent protection in general use are: *fuses* and *circuit breakers*.

Rewirable fuse

A fuse is defined as: 'A device for opening a circuit by means of a conductor, designed to melt when an excessive current flows along it. The fuse comprises all the parts of the complete device'. The rewirable type consists of a porcelain or plastic bridge and a base (carrier) as shown in Fig. 6.14. The bridge has two sets of contacts which fit into other contacts in the base. The fuse-element consists of a fine

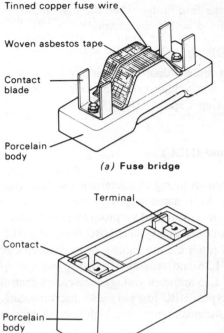

(a) **Fuse bridge**

Tinned copper fuse wire

Woven asbestos tape

Contact blade

Porcelain body

Terminal

Contact

Porcelain body

(b) **Carrier**

Fig. 6.14 The rewirable fuse

tinned copper wire designed to melt and open the circuit when an excess current flows. This is connected between the terminals of the fuse bridge. An asbestos or asbestos substitute tube or pad is provided to reduce the effects of arcing when the fuse-element melts. Three terms are used in connection with fuses.

Current rating   This is the maximum current which the fuse will carry indefinitely without undue deterioration of the fuse element.

Fusing current   This is the minimum current which will 'blow' the fuse.

Fusing factor   This is the ratio of the minimum fusing current to the current rating, namely:

Fusing Factor = (minimum fusing current)/(current rating).

The rewirable fuse is a relatively cheap type of overcurrent protective device and is still widely used despite several disadvantages including:

(a) the ease with which an inexperienced person can replace a 'blown' fuse element with a wire of incorrect thickness (gauge) or type,
(b) undue deterioration of the fuses' elements due to oxidation,
(c) lack of discrimination. This means that it is possible, in certain conditions, for a 15 ampere fuse element to melt before a 10 ampere fuse element. Also, a rewirable fuse is not capable of discriminating between a high current of short duration — such as a motor starting — and a continuous fault current,
(d) damage to the fuse bridge and base in conditions of severe short circuit.

The fusing factor for a rewirable fuse is about 2. Thus a fuse element rated at 10 amperes will melt ('blow') at $10 \times 2 = 20$ amperes, so a fault current of twice the normal design current must flow before the protection device functions and breaks the circuit. This also means that it is possible to run a circuit on a continuous overload for a considerable time with eventual damage to the insulation.

Cartridge fuse

The obvious disadvantages of the rewirable type of fuse led to the development and use of the *cartridge fuse* which is most often found in 13 ampere fused plugs. Figure 6.15 shows the construction of a typical cartridge fuse. The fusing factor for this type of fuse is 1.5. Thus a cartridge fuse rated at 13 amperes will 'blow' at 19.5 amperes. The rating of this type of fuse is determined and fixed by the manufacturer. The advantages of the cartridge fuse are:

(a) the current rating is accurately known,
(b) the fuse element is unlikely to deteriorate in service,
(c) closer control because of its lower fusing factor,

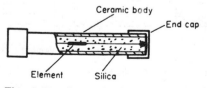

Ceramic body

End cap

Element   Silica

Fig. 6.15 Cartridge fuse

(*d*) there is no damage to the fuse bridge when the fuse 'blows' as any disruption is contained within the cartridge.

The disadvantages of this type of cartridge fuse are:

(*a*) it is more expensive to replace than the rewirable type,
(*b*) it is unsuitable for use where high values of fault current may occur. (See High Breaking Capacity (HBC) fuses.)

High breaking capacity fuse (HBC)

This type of cartridge fuse has its fusing characteristic carefully controlled by the manufacturer. As its name suggests it can safely interrupt very large fault currents. It is often used for protecting main cables. Figure 6.16 shows the construction of a typical HBC fuse. The HBC fuse is more expensive than either the rewirable type or the cartridge type. The fusing factor is 1.25 and thus an HBC fuse rated at 10 amperes will 'blow' at only 12.5 amperes giving much closer control than either of the previous types. HBC fuses are also discriminating. That is, they are able to distinguish between a high starting current

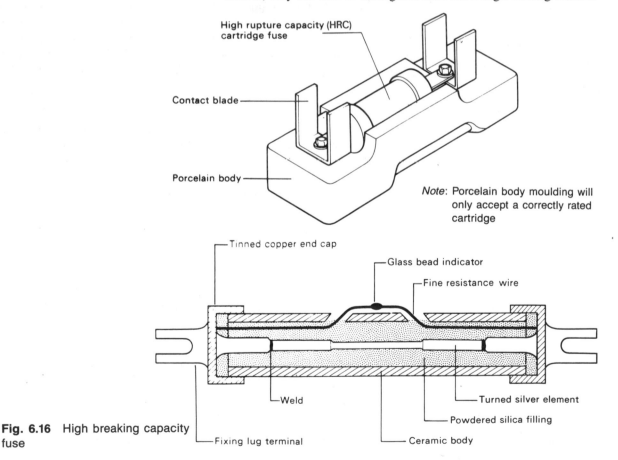

Note: Porcelain body moulding will only accept a correctly rated cartridge

Fig. 6.16 High breaking capacity fuse

taken by a motor (which lasts only a matter of seconds) and a high fault or overload current (which lasts longer).

The type and rating of a fuse must be carefully matched to the circuit it is protecting and care must be taken to replace a 'blown' fuse with one which is of the same type and rating. If the fuse repeatedly 'blows', it must not be replaced with one of increased rating as a fault is obviously present. A qualified electrician should be called in to clear the fault. Note that 13 ampere fused plugs are always sold with a 13 ampere fuse installed. If it is to be fitted to an appliance taking only a small current, such as a table lamp, then a fuse of lower rating must be fitted (3 ampere).

One point that must be remembered about fuses and fuse protection is that the circuit fuse protects the *circuit cables* from being overloaded and should also prevent the main fuse operating in the case of a local short circuit. Circuit fuses *do not protect the appliance* connected to the circuit from fault or overload.

Miniature circuit breakers (MCB)

The usual arrangement is for a spring-loaded contact to be held in place by a gravity-operated trigger. A solenoid in the breaker carries the whole or part of the circuit current and, if this rises above the design maximum for the circuit, the magnetic field produced by the solenoid trips the trigger and the contacts are opened. Figure 6.17 shows

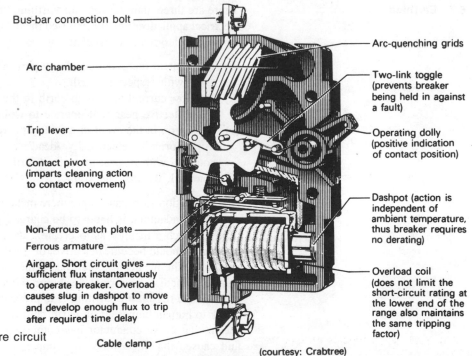

Bus-bar connection bolt

Arc chamber

Trip lever

Contact pivot (imparts cleaning action to contact movement)

Non-ferrous catch plate

Ferrous armature

Airgap. Short circuit gives sufficient flux instantaneously to operate breaker. Overload causes slug in dashpot to move and develop enough flux to trip after required time delay

Cable clamp

Arc-quenching grids

Two-link toggle (prevents breaker being held in against a fault)

Operating dolly (positive indication of contact position)

Dashpot (action is independent of ambient temperature, thus breaker requires no derating)

Overload coil (does not limit the short-circuit rating at the lower end of the range also maintains the same tripping factor)

Fig. 6.17 The miniature circuit breaker

(courtesy: Crabtree)

a typical miniature circuit breaker (MCB). The advantages of this type of overcurrent protection device are that:

(*a*) it gives closer protection than any fuse device,
(*b*) fault current instantly disconnects the supply,
(*c*) fault current tripping characteristics are set by the manufacturer and cannot be altered,
(*d*) breaker will trip for small, sustained overloads but not for harmless surge currents (transients),
(*e*) the faulty circuit is indicated by the position of the manual re-set lever,
(*f*) supply is quickly and easily restored by raising the re-set lever (dolly) when the fault has been cleared.

However, miniature circuit breakers also have quite a number of disadvantages compared with the more simple fuses and these have, so far, precluded their universal adoption. These disadvantages are that:

(*a*) they are very much more expensive than a fuse of the same rating,
(*b*) they are very much more bulky than a fuse of the same rating, particularly in the smaller current ratings,
(*c*) they must be regularly tested to ensure that they will operate correctly when a fault occurs,
(*d*) the mechanism becomes sluggish and unreliable if left unused for long periods.

6.7 Earthing

There are three main reasons for earthing the exposed metalwork of electrical appliances and the exposed metalwork of buildings in which an electrical system is installed.

(*a*) To maintain the potential of any part of the system at a definite value with respect to earth.
(*b*) To allow current to flow to earth in the event of a fault, so that the protective gear will operate to isolate the faulty circuit.
(*c*) To ensure, in the event of a fault, apparatus and exposed metalwork normally electrically 'dead', cannot reach a dangerous potential with respect to earth. (Earth is assumed to be at zero potential, 0 volts, 'no volts'.)

IEE Regulation 13-8 states that where metalwork, other than current-carrying conductors, is liable to be charged with electricity (become 'live') in such a manner as to create a danger if the insulation of a conductor should become defective; or if a defect should occur in any apparatus, the metalwork shall be earthed in such a manner as will ensure immediate electrical discharge without danger.

The basic reason for earthing is to prevent or minimise the risk of shock to human beings. If an earth fault occurs in an installation, it means that a live conductor has come into contact with metalwork and caused that metalwork to become 'live', that is, to cause the metalwork to reach the same potential as the conductor. Any person touching the metalwork whilst standing on a non-insulating floor, or touching earthed metalwork at the same time, will receive an electric shock as the fault current passes through their body to earth. If, however, the metalwork is connected to the general mass of the earth through a very low-resistance path, the fault current will flow to earth through the low resistance earth conductor rather than through the very high resistance of the human body. If the fault current is sufficiently heavy it will then 'blow' a fuse and isolate the faulty circuit. Figure 6.18(*a*) shows the *earth-fault-loop path* and the way in which a fault current blows the protective fuse and isolates the circuit. Figure 6.18(*b*) shows the need for earthing metal clad equipment.

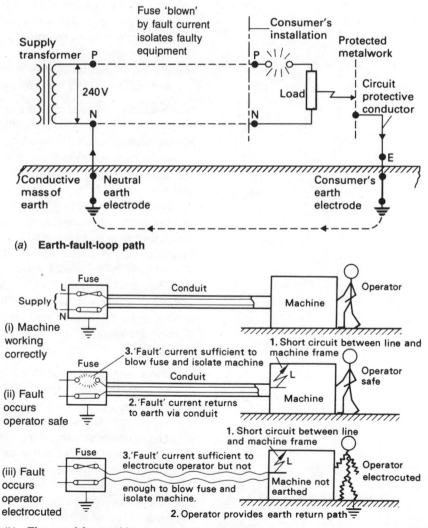

(*a*) **Earth-fault-loop path**

(*b*) **The need for earthing metal clad equipment**

Fig. 6.18 The need for earthing

However, an earth fault-current may flow to earth with insufficient magnitude to 'blow' the fuse, yet be sufficient to cause a fatal 'shock' or cause overheating at, say, a loose connection and to start a fire.

Chapter 4.1 of the IEE Regulations deals with the requirements which all earthing arrangements must satisfy if an electrical installation is to be deemed safe. The basic requirements are:

(a) the complete insulation of all parts of an electrical system. This involves the use of apparatus of 'all-insulated' construction, which means that the insulation which encloses the apparatus is durable and substantially continuous,

(b) the use of appliances with *double insulation* conforming to British Standard Specifications,

(c) the earthing of exposed metal parts (there are some exemptions) not only of apparatus connected to the supply, but of all exposed metalwork within the building served by the supply, e.g. water and gas pipes,

(d) the isolation of metalwork in such a manner that it is not liable to come into contact with any live parts or with earthed metalwork.

Good earthing requires that the earthing arrangements of the consumer's installation are such that the occurrence of a fault of negligible impedance from a live conductor to adjacent, exposed metal will cause a current which is three times the fuse rating or one and one-half times the setting of the overcurrent circuit-breaker to flow. This will immediately 'blow' the fuse or trip the circit-breaker and render the faulty circuit 'dead'. The earthing arrangements should be such that the maximum sustained voltage developed under fault conditions between exposed metal required to be earthed and the consumer's earth terminal shall not exceed 50 volts. The 'earth wire' connecting (bonding) together all metalwork required to be earthed is called the *Circuit Protective Conductor* (CPC). If it is a separate conductor (insulated and coloured green or green and yellow) it must have a cross-sectional area of at least 1.00 mm^2 and need not be greater than 70 mm^2.

There are a number of methods used to achieve the earthing of an installation:

(a) connection to the metal sheath and armouring of a supply authority's underground supply cable,

(b) connection to the continuous earth wire (CEW) provided by a supply authority where distribution of energy is by overhead lines,

(c) connection to an earth electrode sunk into the ground for the purpose. Note that cold-water supply pipes no longer form a safe and efficient earth return path since, increasingly, plastic piping is being used for this purpose,

(d) installation of a protective-multiple earthing system (PME),

(e) installation of automatic fault protection.

Connection to an earth electrode driven into the ground is the most common method of providing an earth return path. The buried length of the rod is more important than its diameter and it is the length in contact with the subsoil which is important, contact with the top-soil being disregarded. The PME system (Regulation Appendix 3) gives protection against earth-fault conditions and uses the neutral of the incoming supply as the earth point or terminal. In this system of earthing, all protected metalwork is connected, by means of the installation CPCs, to the neutral service-conductor at the supply intake position, thus forming a 'Faraday cage'. By doing this, line-to-earth faults are converted into line-to-neutral faults in order to ensure that sufficient fault-current will flow to 'blow' a fuse or trip an overload circuit-breaker and isolate the fault from the supply. However this system has a number of disadvantages and stringent requirements are laid down to cover the use of the system. Figure 6.19 shows a typical distribution system with a number of consumers connected to a common PME system of earthing.

The installation of automatic earth-fault protection equipment has increased in recent years. This involves the use of *earth leakage circuit-breakers* (ELCB). There are two methods of protection.

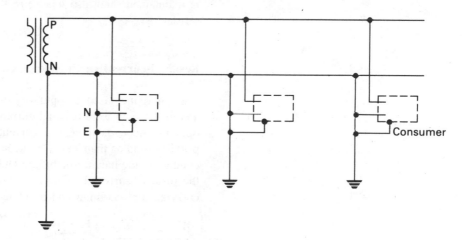

Fig. 6.19 Protective multiple earthing (PME) system

Fault-voltage earth-leakage circuit-breaker

Fault-voltage units are designed to be directly responsive to fault voltages which appear on protected metalwork, which happens when the earth-path resistance is relatively high. They operate on leakage currents as low as 50 milli-amperes. Figure 6.20 shows the circuit for a typical ELCB connected to a protected installation. Note that the trip coil is connected between the protected metalwork and earth in the same way as a voltmeter would be connected to measure the potential between the metalwork and earth. If a fault occurs, the fault-current will flow through the trip coil to energise it and so trip the breaker contacts. The test switch is provided so that the unit can be tested at frequent intervals to make sure that the continuity of the earth path

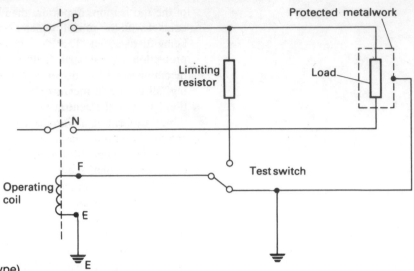

Fig. 6.20 ELCB (fault-voltage type)

is maintained, and also to ensure that the operation of the unit is satisfactory.

Residual-current earth-leakage circuit-breaker

These have increased in popularity in recent years. Figure 6.21 shows a typical circuit for a residual current ELCB. The basic principle of operation depends upon more current flowing into the live side of the primary winding than leaves by the neutral conductor and/or the earth conductor, the balance being the fault-current. The essential part of the residual-current ELCB is a transformer with opposed windings carrying the incoming and outgoing currents. In a healthy circuit,

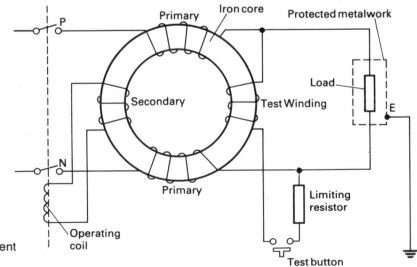

Fig. 6.21 ELCB (residual-current type)

where the values of the currents in the windings are equal, the magnetic effects cancel each other out. However, a fault will cause an out-of-balance condition and create a magnetic effect in the transformer core which links with the turns of a small secondary winding. This secondary winding is permanently connected to the trip coil of the circuit-breaker. Thus the current induced in the secondary winding by the out-of-balance condition will energise the trip coil and open the circuit-breaker, disconnecting the supply. A test winding is also provided and current can be passed through it by operation of the test switch. Operation of the test circuit unbalances the transformer and should cause the circuit-breaker to trip if the ELCB is functioning correctly.

Portable electric tools should be operated at reduced potential (110 volts). This is provided by a step-down transformer with separate primary and secondary windings. The secondary windings and the metalwork of the transformer should be connected to earth and the transformer secondary winding is also protected from overcurrent faults by suitable fuses. The primary winding should be connected to the mains via an ELCB.

6.9 Measuring instruments

The testing of electrical circuits involves the measurement of:

(a) potential (volts),
(b) current (amperes),
(c) insulation resistance and circuit resistance (ohms),
(d) continuity (no units).

The instruments most frequently used for such tests are:

(a) continuity tester (lamp/buzzer and battery),
(b) multi-range meter (volts, amps and ohms),
(c) 'megger' resistance and continuity tester.

Figure 6.22(a) shows the circuit for a simple continuity tester. It can be used where the resistance of the cables is relatively low and the distances run are short, as in the majority of domestic and office lighting and power circuits. If the circuit is complete, then the buzzer will sound and the lamp will light up. Note that when testing the distribution fuse board, it is isolated by the main switch and the fuse protecting the final circuit under test is removed before the continuity test is made. This isolation of the circuit is for safety, and also to ensure that the test is only applied to one circuit at a time as shown in Fig. 6.22(b). The appliance completing the circuit is bridged in order to reduce the resistance of the circuit before the test is made. The continuity tester can also be used to identify wires as shown in Fig. 6.22(c). Where a long run of wiring is under test the volt-drop may be too great for the battery operated continuity tester shown in Fig. 6.22. In this case a 'megger' insulation and continuity tester is used. Such an instrument is shown in Fig. 6.23.

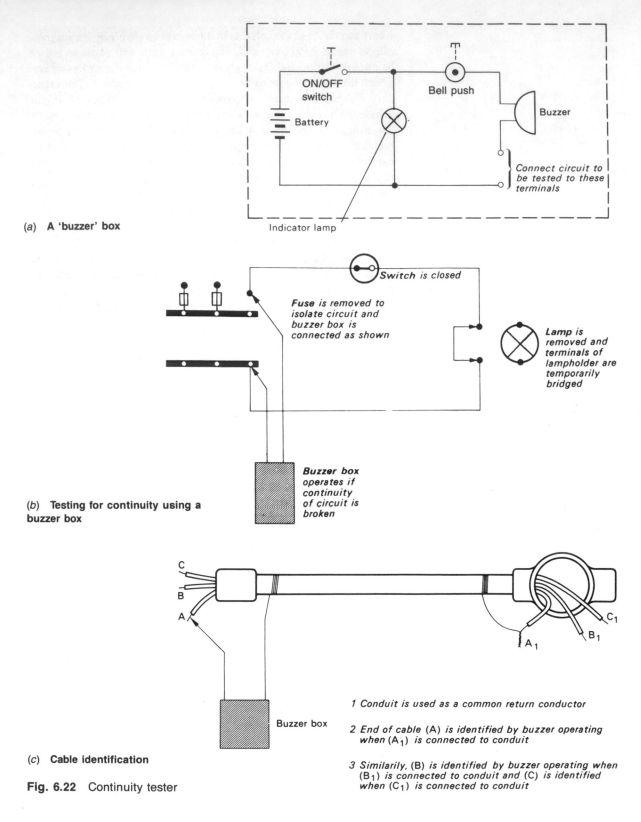

(a) **A 'buzzer' box**

(b) **Testing for continuity using a buzzer box**

(c) **Cable identification**

Fig. 6.22 Continuity tester

Connect circuit to be tested to these terminals

Indicator lamp

Switch is closed

Fuse is removed to isolate circuit and buzzer box is connected as shown

Lamp is removed and terminals of lampholder are temporarily bridged

Buzzer box operates if continuity of circuit is broken

1 *Conduit is used as a common return conductor*

2 *End of cable (A) is identified by buzzer operating when (A$_1$) is connected to conduit*

3 *Similarily, (B) is identified by buzzer operating when (B$_1$) is connected to conduit and (C) is identified when (C$_1$) is connected to conduit*

	WM5/500
Insulation Test Voltage	500 V d.c.
Ranges insulation continuity	0–200 MΩ 0–100 Ω

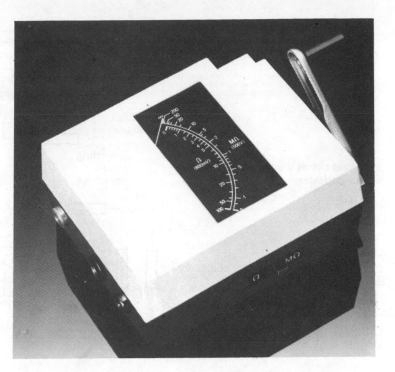

Fig. 6.23 The Megger

The 'megger'

This instrument consists of an ohm-meter (resistance meter) and a hand-driven generator combined together in a single case. The circuit diagram of a 'megger' is shown in Fig. 6.24(*a*) when switched for insulation-resistance testing, and in 6.24(*b*) when switched for continuity testing. This instrument is similar to a moving coil meter except that it has two coils mounted at right angles to each other on its spindle. The control coil is connected across the generator and the deflecting coil is connected in series with the circuit under test. When connected to a circuit of infinite resistance ('open' circuit) only the control coil is energised and the pointer swings across the dial to infinity. When a 'closed' circuit is connected across the 'megger', current will flow through the deflecting coil. The force produced by the deflecting coil opposes that of the control coil and the needle will take up some intermediate position. In the event of a short circuit being measured the deflecting coil will overwhelm the control coil and the instrument will read zero resistance.

The requirements of insulation resistance tests for new installations are laid down in IEE Regulations, Part 6. Figure 6.25 shows the correct method for testing insulation resistance with respect to earth. Figure 6.26 shows the correct method for testing insulation between conductors. Figure 6.27 shows the correct method for testing the insulation and continuity of appliances. All these examples refer to domestic installations fed from a two wire service main.

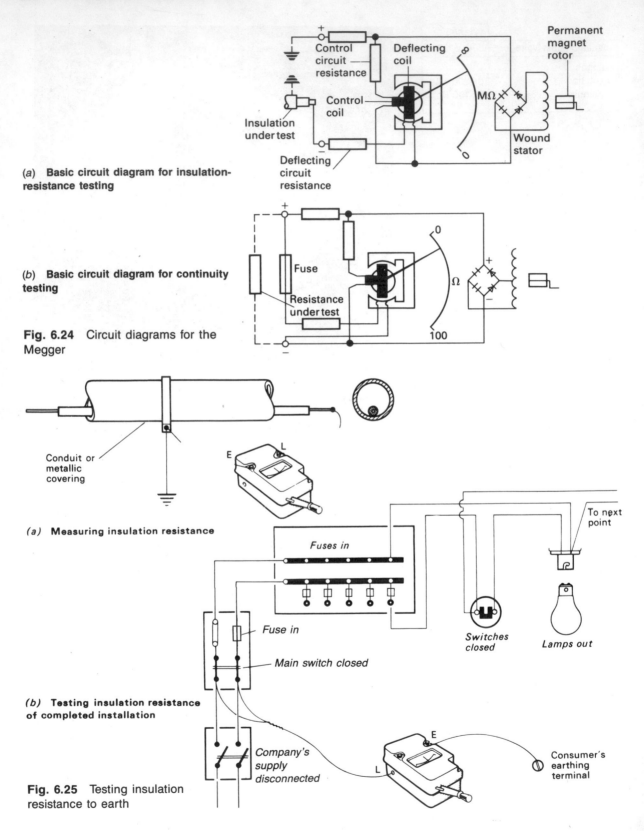

(a) **Basic circuit diagram for insulation-resistance testing**

(b) **Basic circuit diagram for continuity testing**

Fig. 6.24 Circuit diagrams for the Megger

(a) **Measuring insulation resistance**

(b) **Testing insulation resistance of completed installation**

Fig. 6.25 Testing insulation resistance to earth

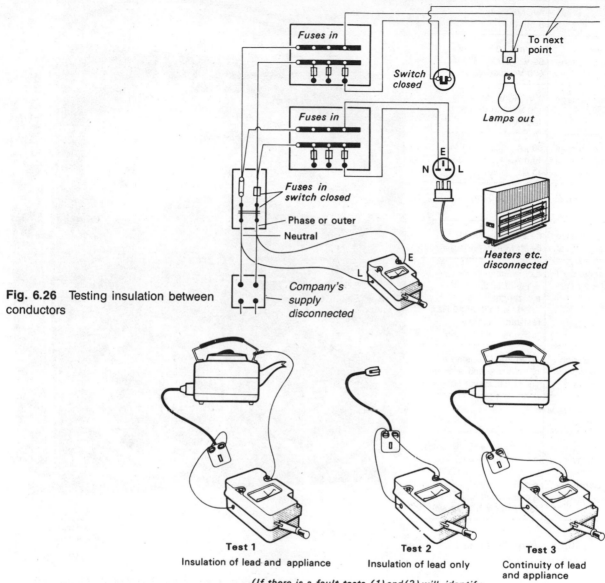

Fig. 6.26 Testing insulation between conductors

Fig. 6.27 Testing an appliance

Test 1
Insulation of lead and appliance

Test 2
Insulation of lead only

Test 3
Continuity of lead and appliance

(If there is a fault tests (1) and (2) will identify whether it is in the lead or the appliance)

The multi-range meter

The multi-range meter is designed to read potential (volts), current (amperes), and resistance (ohms). Figure 6.28 shows a typical instrument and its ranges. The basis of this instrument is a sensitive moving coil meter and switching is provided so that various values of voltage multiplier resistance can be connected in series with the meter depending upon the voltage being measured, or various values of shunt resistance can be connected in parallel with the meter depending upon the current being measured. A bridge rectifier circuit is provided when

	AVOMETER 8 mk 6
Ranges d.c. voltage	100 mV 3 V 10 V 30 V 100 V 300 V 600 V 1000 V
a.c. voltage	3 V 10 V 30 V 100 V 300 V 600 V 1000 V
d.c. current	50 μA 300 μA 1 mA 10 mA 100 mA 1 A 10 A
a.c. current	10 mA 100 mA 1 A 10 A
Resistance	0–2 kΩ 0–200 kΩ 0–20 MΩ
Accuracy (at 20°C)	d.c. ranges ±1% of f.s.d. a.c. ranges ±2% of f.s.d. at 50 Hz resistance ranges ±5% at centre scale
Insulation Resistance Range	up to 200 MΩ using ohms scale and external 150 V d.c. power supply
Capacitance Range	
Decibels Range	−10 dB to +55 dB using a.c. voltage scale
Sensitivity	d.c. 20 kΩ/volt a.c. 2 kΩ/volt (30 V range and above)

Fig. 6.28 A multi-range test meter

measuring a.c. voltages. The instrument shown is an *analogue* meter since it uses a pointer and scale. *Digital* meters give a direct numerical reading. The latter type of meter, although more expensive, has the advantages of being easier to read, and more mechanically robust (no sensitive movement to jolt off its pivots). However it requires an internal battery which can always run down at inopportune moments and, if the test voltage or current is not stable, the display can flutter in a confusing manner. Figure 6.29 shows how the meter is connected to read potential and current. Although it can be used to measure insulation resistance and continuity, it must be remembered that these tests are only being carried out at the low potential of the internal battery and are, therefore, no substitute for the 'megger' insulation test.

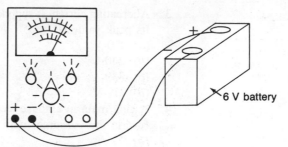

(*a*) **Multi-range meter measuring EMF (parallel connection)**

Meter set to 0–15 V d.c. range

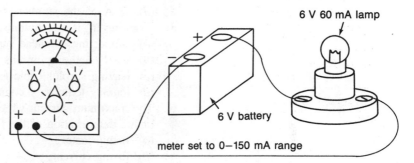

(*b*) **Multi-range meter measuring current (series connection)**

6 V 60 mA lamp

6 V battery

meter set to 0–150 mA range

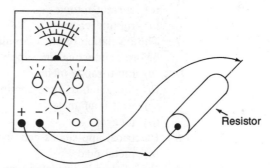

(*c*) **Multi-range meter measuring resistance (parallel connection)**

Resistor

Fig. 6.29 Uses of a multi-range meter

Meter switched to appropriate resistance range
Ohmic value is read directly off meter scale

Problems

Section A

1 The lights in a house are normally wired in:
   (*a*) series;
   (*b*) parallel;
   (*c*) neither;
   (*d*) both.

2 Which of the following items of equipment is *not* provided by the electricity supply company:
   (*a*) service cable;
   (*b*) service fuse and neutral link;
   (*c*) meter;
   (*d*) consumer unit.

3   Alternating current is supplied to a house at a potential of 240 V r.m.s. Its peak value is:
    (a) 170 V;
    (b) 240 V;
    (c) 339 V;
    (d) 415 V.

4   A ring main in a house feeds a number of 13 A socket outlets. The fuse in the consumer unit feeding the ring main will be rated at:
    (a) 13 A;
    (b) 20 A;
    (c) 30 A;
    (d) 13 A × the number of socket outlets.

5   'Looping-in' is a technique for:
    (a) supporting overhead supply cables;
    (b) wiring lighting circuits in parallel;
    (c) leaving surplus cable for future extensions;
    (d) making temporary connections;

6   The maximum current which a fuse will carry indefinitely without undue deterioration of the fuse element is called the:
    (a) fusing factor;
    (b) fusing current;
    (c) current rating;
    (d) rupture capacity.

7   The insulation of each circuit of a new installation should be tested before connecting the supply, using a:
    (a) multi-range meter;
    (b) 'buzzer box' type continuity tester;
    (c) watt-hour meter;
    (d) megger.

8   Electrical appliances need not be earthed if they are:
    (a) metal clad;
    (b) of double insulated construction;
    (c) used in a room with plastic floor tiles;
    (d) rated at less than 1 kW.

9   Which of the following devices is becoming increasingly popular for earth fault protection:
    (a) ELCB;
    (b) MCB;
    (c) rewirable fuse;
    (d) cartridge fuse.

10 The IEE Regulations state that every connection or joint shall be:
    (a) electrically sound;
    (b) mechanically sound;
    (c) electrically sound and insulated;
    (d) mechanically and electrically sound.

Section B

11 With the aid of sketches explain the difference between a ring circuit and a radial circuit and give an example where each would be used.

12 Explain what is meant by 'Diversity Factor' and how it is applied to the lighting circuits in private houses. State the permitted factor for lighting circuits in a private house.

13 Compare the advantages and limitations of fuses and miniature circuit breakers for circuit protection against overcurrent faults.

14 (*a*) With the aid of a diagram explain why metal clad equipment should be earthed.

    (*b*) With the aid of a diagram explain what is meant by a 'Protective Multiple Earth' system.

15 A typical modern house has a consumer unit with the following outlets.

    (i) Three fuses rated at 30 amperes each.

    (ii) One fuse rated at 15 amperes.

    (iii) Two fuses rated at 5 amperes each.

    State which circuit you would connect to each fuse giving reasons for your choice.

16 (*a*) With the aid of a diagram explain the construction and function of a 'megger'.

    (*b*) With the aid of diagrams, show two tests which can be made with this instrument.

17 With the aid of diagrams explain three typical tests which can be performed with a multi-range meter.

18 Draw a circuit diagram, using symbols from BS 3939, showing 3 lighting points wired in parallel and looped in at the ceiling roses, together with a transformer fed trembler bell circuit operated by a bell-push, all fed from the same fuse in the consumer unit.

19 Explain why a fuse is fitted in a 13 ampere plug, why the rating of the fuse can be varied, and which circuit this fuse protects.

20 Discuss the problems associated with the switching and overcurrent protection of inductive circuits such as fluorescent lights and electric motors. For example, why are 'slow-burn' fuses often used to protect such circuits.

7 Electronic principles and practice

7.1 Introduction

Electronics can be defined broadly as that branch of electrical engineering which exploits the properties of electrons in controlling the flow of current, rather than by using electro-mechanical switching devices.

Figure 7.1 shows a typical electronic circuit. In this example it is a small signal amplifier. It can be seen that the circuit consists of such components as resistors, capacitors, inductors (chokes) and a transformer together with a transistor and diodes. Later in this chapter actual values will be given to this circuit so that it can be constructed and tested.

The resistors, capacitors and inductors are all used to control the flow of current, but they are all incapable of providing gain (amplifica-

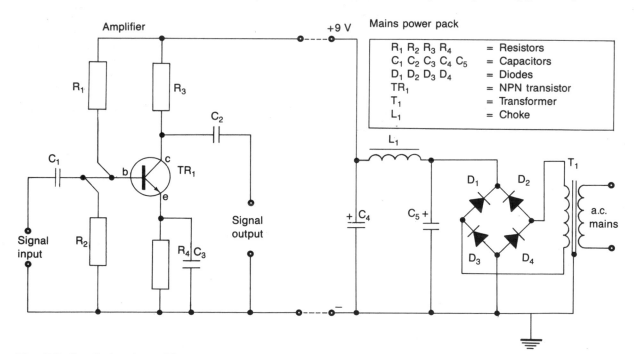

Fig. 7.1 Small signal amplifier

tion). Therefore they are referred to as *passive* components. The transistor, on the other hand, does provide gain (amplification). Therefore it is referred to as an *active* component. In order to provide this gain the transistor has to draw additional energy from a battery or, as shown in Fig. 7.1, a mains operated power-pack. These components, and variations upon them, will now be considered in greater detail.

7.2 Resistors

Resistors are passive devices which control the flow of current in electronic circuits. The theory of electrical resistance in conductors, including Ohm's law, resistivity, and the characteristics of series and parallel circuits is dealt with in depth in your Physical Science unit. The theory of resistance in electrical circuits is summarised in Fig. 7.2. Resistors used in electronic circuits may be fixed in value or variable.

7.3 Resistor identification

The non-inductive resistors, that is, carbon compound, carbon film, metal film and metal oxide, are usually identified by colour code or by a number and letter code. Figure 7.3 shows how the colour code is applied. In this example:

Band 1 is *orange* = 3
Band 2 is *violet* = 7
Band 3 is *green* = 5 That is, a multiplying factor of 10^5

Thus the value of the resistance is $37 \times 10^5 = 3\ 700\ 000$ ohm or 3.7 megohm. More simply, the number related to the colour of band 3 can be considered as the number of zeros following the first two numbers. This gives the same answer.

The fourth band indicates the manufacturing tolerance. For the example shown the silver band indicates a 10% tolerance. That is, $3\ 700\ 000\ \Omega \pm 10\%$. Therefore the resistance of this resistor can vary between 370 000 Ω larger than the nominal value, or 370 000 Ω smaller than the nominal value. So when you use a 3 700 000 Ω resistor with a 10% tolerance, its value can lie anywhere between 3.33 MΩ and 4.07 MΩ. This is a very big variation, but a resistor with no tolerance band would vary twice as much as this, whilst one with a gold tolerance band would vary half as much. This is the reason why carbon compound resistors are only used where cheapness is of prime importance and for applications where accuracy does not matter. Where greater accuracy is required, the high stability resistors are used and these are available with tolerances as close as 1%.

The number and letter code recommended by BS 1852 is shown in Table 7.1. Although introduced in 1976, it is difficult to read on the smaller sizes of non-inductive resistors. Hence the continued use of the traditional colour code. Wire wound resistors are physically larger and there is no problem in using this code with them.

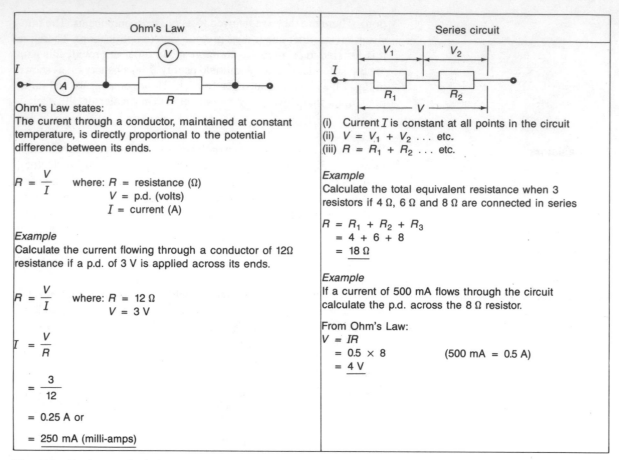

Ohm's Law	Series circuit

Ohm's Law

Ohm's Law states:
The current through a conductor, maintained at constant temperature, is directly proportional to the potential difference between its ends.

$$R = \frac{V}{I}$$ where: R = resistance (Ω)
V = p.d. (volts)
I = current (A)

Example
Calculate the current flowing through a conductor of 12Ω resistance if a p.d. of 3 V is applied across its ends.

$$R = \frac{V}{I}$$ where: R = 12 Ω
V = 3 V

$$I = \frac{V}{R}$$

$$= \frac{3}{12}$$

$$= 0.25 \text{ A or}$$

$$= 250 \text{ mA (milli-amps)}$$

Series circuit

(i)   Current I is constant at all points in the circuit
(ii) $V = V_1 + V_2$ ... etc.
(iii) $R = R_1 + R_2$ ... etc.

Example
Calculate the total equivalent resistance when 3 resistors if 4 Ω, 6 Ω and 8 Ω are connected in series

$$R = R_1 + R_2 + R_3$$
$$= 4 + 6 + 8$$
$$= 18 \ \Omega$$

Example
If a current of 500 mA flows through the circuit calculate the p.d. across the 8 Ω resistor.

From Ohm's Law:
$$V = IR$$
$$= 0.5 \times 8 \qquad (500 \text{ mA} = 0.5 \text{ A})$$
$$= 4 \text{ V}$$

Fig. 7.2   Resistor calculations

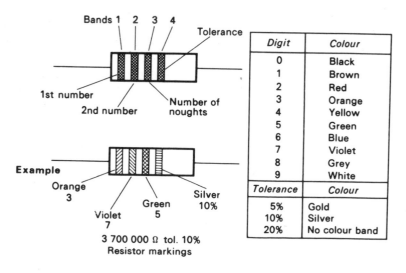

Digit	Colour
0	Black
1	Brown
2	Red
3	Orange
4	Yellow
5	Green
6	Blue
7	Violet
8	Grey
9	White

Tolerance	Colour
5%	Gold
10%	Silver
20%	No colour band

3 700 000 Ω tol. 10%
Resistor markings

Fig. 7.3   Carbon resistors colour code

Note: small capacitors are marked in the same manner, but the units are picofarads

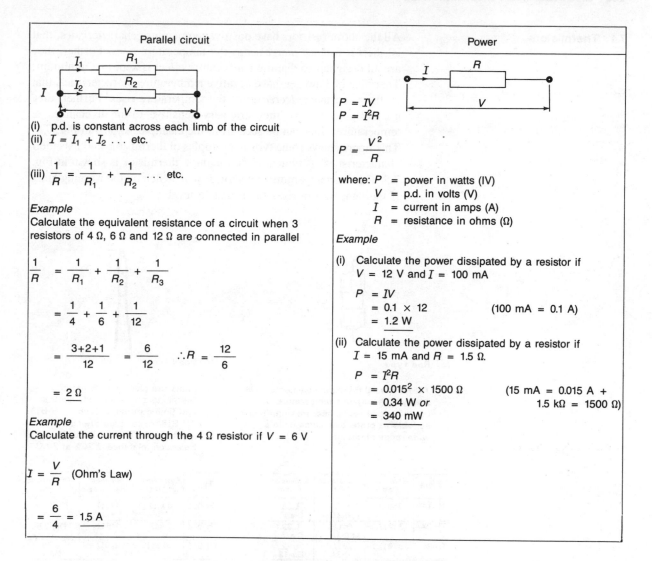

Parallel circuit

(i) p.d. is constant across each limb of the circuit

(ii) $I = I_1 + I_2$ ... etc.

(iii) $\dfrac{1}{R} = \dfrac{1}{R_1} + \dfrac{1}{R_2}$ ... etc.

Example

Calculate the equivalent resistance of a circuit when 3 resistors of 4 Ω, 6 Ω and 12 Ω are connected in parallel

$$\frac{1}{R} = \frac{1}{R_1} + \frac{1}{R_2} + \frac{1}{R_3}$$

$$= \frac{1}{4} + \frac{1}{6} + \frac{1}{12}$$

$$= \frac{3+2+1}{12} = \frac{6}{12} \quad \therefore R = \frac{12}{6}$$

$$= 2\,\Omega$$

Example

Calculate the current through the 4 Ω resistor if $V = 6$ V

$$I = \frac{V}{R} \quad \text{(Ohm's Law)}$$

$$= \frac{6}{4} = 1.5\ \text{A}$$

Power

$P = IV$

$P = I^2R$

$$P = \frac{V^2}{R}$$

where: P = power in watts (IV)
  V = p.d. in volts (V)
  I = current in amps (A)
  R = resistance in ohms (Ω)

Example

(i) Calculate the power dissipated by a resistor if $V = 12$ V and $I = 100$ mA

  $P = IV$
    $= 0.1 \times 12$ (100 mA = 0.1 A)
    $= 1.2$ W

(ii) Calculate the power dissipated by a resistor if $I = 15$ mA and $R = 1.5$ Ω.

  $P = I^2R$
    $= 0.015^2 \times 1500\ \Omega$ (15 mA = 0.015 A +
    $= 0.34$ W *or*   1.5 kΩ = 1500 Ω)
    $= 340$ mW

Table 7.1 Number and letter code for resistors (BS 1852)

The British Standard number and letter code for small resistors is now replacing the traditional colour code:

0.47 Ω would be marked R47	100 Ω would be marked 100R
1 Ω would be marked 1R0	1 kΩ would be marked 1K0
4.7 Ω would be marked 4R7	10 kΩ would be marked 10K
47 Ω would be marked 47R	10 MΩ would be marked 10M

After this code is added a letter to indicate tolerance:

F = ±1%   J = ±5%   M = ±20%
G = ±2%   K = ±10%

Thus

  4M7M = 4.7 MΩ ± 20%

7.4 Thermistors

All the above resistors have positive temperature characteristics, that is, their resistance *increases* as the temperature rises. Further, they are all designed so that the temperature effect is kept to a minimum. Thermistors, however, have negative temperature characteristics, that is, their resistance *decreases* as the temperature rises. Further, they are designed so that they are very sensitive to small changes in temperature. This can be seen from the examples shown in Fig. 7.4. This figure shows some typical examples of thermistors and lists their characteristics. A typical circuit using a thermistor is shown in Fig. 7.5. This is a temperature detector. It will switch on the warning lamp if the temperature rises to an unsafe level.

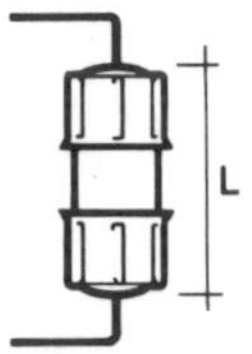

(a) **Rod type**

These special resistance elements have a very high negative temperature coefficient of resistance, making them suitable as protective elements in a wide range of circuits

(b) **Bead type**

Miniature glass-encapsulated thermistors for amplitude control and timing purposes (Types TH-B15, TH-B18) or temperature measurement (Types TH-B11, TH-B12). Selection tolerance ±20% at 20 C

Type	*Resistance Cold*	*Resistance Hot*	*Dimensions in mm*
TH-1A	650 Ω	37 Ω at 0·3A	L.38 Dia.11
TH-2A	3·8kΩ	44 Ω at 0·3A	L.32 Dia. 8
TH-3	370Ω	28 Ω at 0·3A	L.22 Dia.12
TH-5	4 Ω	0·4 Ω at 1W (max)	Dia.10 H.4·5

Type	*Resistance at 20 C*	*Minimum resistance*	*Dimensions in mm*
TH-B11	1 MΩ	170 Ω	L.10 Dia 2·5
TH-B12	2 kΩ	115 Ω	
TH-B15	100 kΩ	320 Ω	L.25 Dia.4
TH-B18	5 kΩ	100 Ω	L.38 Dia.10

Fig. 7.4 Thermistors

7.5 Capacitors

These are devices for storing electrical charges. Unlike secondary cells they can charge and discharge very rapidly. Figure 7.6 shows the construction of a simple capacitor and its circuit symbol. Figure 7.7(*a*) shows a capacitor connected across a battery. At the moment of switch-on, current will flow as negative electrons move from the negative pole of the cell to plate 'A' of the capacitor. Since like charges repel each other, the build up of the negative charge on plate 'A' will repel the electrons on plate 'B' which are, in any case, attracted towards

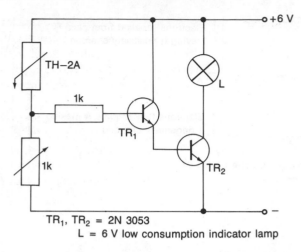

Fig. 7.5 Temperature sensitive circuit

TR$_1$, TR$_2$ = 2N 3053
L = 6 V low consumption indicator lamp

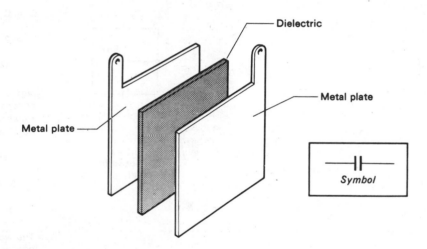

Fig. 7.6 A simple capacitor (exploded view)

the positive pole of the battery. The deficiency of electrons on plate 'B' gives it a positive charge. This process continues until the potential across the plates of the capacitor equals that of the battery. The current flow is not constant. It is rapid at first and then slows down as the potential difference of the capacitor approaches that of the battery and this is shown in Fig. 7.7(*b*).

The storage capacity of a capacitor can be increased by increasing the area of the plates or decreasing the thickness of the insulation between them (the dielectric). The limiting factor for dielectric thickness is the point when the electrical potential applied across the plates breaks down the dielectric insulation and a 'flash-over' occurs. The storage capacity of a capacitor is called its *capacitance*. The unit of capacitance

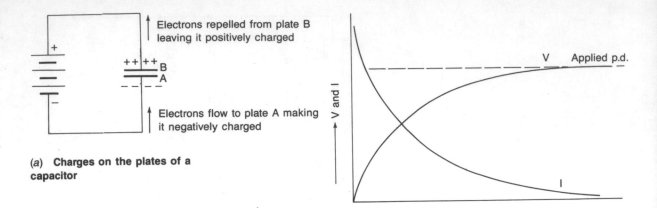

Electrons repelled from plate B
leaving it positively charged

Electrons flow to plate A making
it negatively charged

(a) **Charges on the plates of a
capacitor**

Fig. 7.7 Charging a capacitor

(b) **Charging cycle for a capacitor (d.c.)**

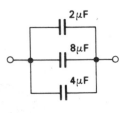

$$C = C_1 + C_2 + C_3$$

$$= 2 + 8 + 4$$

$$= 14\,\mu F$$

Note: *the working voltage of a number
of capacitors connected in parallel is
equal to the capacitor with the lowest
working voltage, e.g.,*

(a) **Capacitors
connected in
parallel**

$2\,\mu F$, 350V d.c. working

$8\,\mu F$, 600V d.c. working

$4\,\mu F$, 600V d.c. working

The working voltage for the above capacitors
connected in parallel is 350V

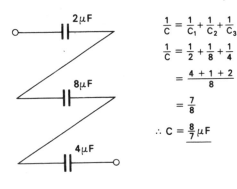

$$\frac{1}{C} = \frac{1}{C_1} + \frac{1}{C_2} + \frac{1}{C_3}$$

$$\frac{1}{C} = \frac{1}{2} + \frac{1}{8} + \frac{1}{4}$$

$$= \frac{4 + 1 + 2}{8}$$

$$= \frac{7}{8}$$

$$\therefore C = \frac{8}{7}\,\mu F$$

Note
Working voltage for series connection is *not* the sum of
the working voltages of the individual capacitors unless
they are identical in all respects

(b) **Capacitors
connected in
series**

Fig. 7.8 Capacitor calculations

is the *farad* (F). It is defined as follows:

a capacitor has a capacitance of 1 farad when a potential difference of 1 volt is required to maintain a charge of 1 coulomb.

$$Q = CV$$

where: Q = charge in coulombs
C = capacitance in farads
V = the p.d. across the capacitor in volts.

1 coulomb represents a charge of 6.3×10^{18} electrons or, 1 coulomb is the charge which flows in one second past any point in a circuit in which there is a steady current of 1 ampere.

The farad (F) is too large a unit for normal purposes and smaller units are used in practice.

1 micro-farad (μF) = 1×10^{-6} F = 0.000 001 farad (F)
1 nano-farad (ηF) = 1×10^{-9} F = 0.000 000 001 farad (F)
1 pico-farad (ρF) = 1×10^{-12} F = 0.000 000 000 001 farad (F)

Like resistors, capacitors can be connected in series and parallel. Figure 7.8 shows how the circuit values for the parallel and series connection of capacitors can be calculated. The ability of a capacitor to store a charge of high voltage electrical energy and to discharge itself almost instantaneously makes it potentially dangerous. It can cause lethal shocks and start fires. High capacitance capacitors used on high voltage supplies must always have a bleed resistance connected across them as shown in Fig. 7.9 to discharge the capacitor when switched off. A suitable value for the bleed resistor is 100 kΩ with a power rating of 3 W.

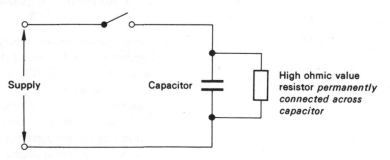

Fig. 7.9 Automatic discharge of a capacitor

7.6 Time constant

One very important circuit combining a resistor and a capacitor in series is shown in Fig. 7.10(*a*). This is used in many timing devices and the *time constant* can be controlled by selecting suitable values of R and C. When the circuit is completed at switch-on, the current I is a maximum since there is no charge on C. The only factor limiting the magnitude of the current flow is R. As V_C approaches V_S, current I approaches zero. Figure 7.10(*b*) shows the relationship between V_C and V_S immediately after 'switch on'.

The *time constant* (the Greek letter 'tau' (τ or T) for a resistance-

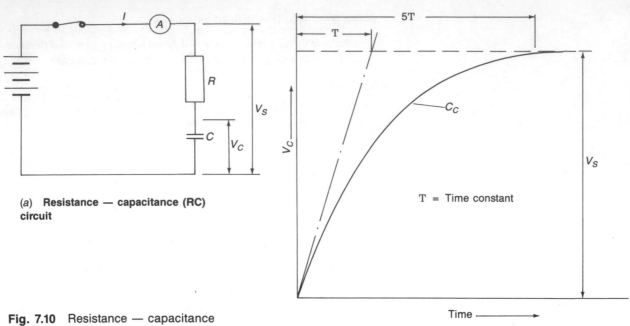

(a) **Resistance — capacitance (RC) circuit**

Fig. 7.10 Resistance — capacitance (RC) circuit

(b) **p.d. across capacitor**

capacitance circuit is defined as the time that would be needed for the p.d. across the capacitor to reach its final steady state value if it had continued at its original rate of increase at switch-on. For all practical purposes the time constant is considered to be the time taken for V_C to reach 63% of V_S.

$$T \text{ (seconds)} = C \text{ (farads)} \times R \text{ (ohms)}.$$
$$T = CR$$

It is not possible to determine exactly when V_C exactly equals V_S and I becomes zero, but for all practical purposes it is assumed that this occurs in 5T seconds (i.e. five time constants). Figure 7.11 shows two worked examples.

So far, capacitors have only been considered in relationship with direct current. When connected into an alternating current circuit a continuous current appears to flow through the capacitor. In fact current can never flow *through* a capacitor because of its insulating dielectric. However, since alternating current flows first in one direction and then in the other, the capacitor is constantly charging and discharging and the current flowing in the circuit will be the charging and discharging currents of the capacitor. The bigger the capacitance of the capacitor, the bigger will be these currents. Consider a lamp in series with a capacitor of about 8 μF and connected to a d.c. supply. At switch-on the lamp will light for a moment as the capacitor charges up. It then goes out once the capacitor is fully charged and no further current flows. If the lamp and capacitor are fed from mains a.c., the lamp will glow continuously as the capacitor charges and discharges

Example 1
Find the time taken for the R.C. circuit shown to reach steady state conditions (5T),

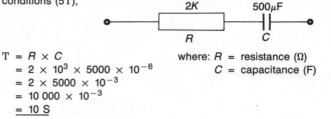

$$T = R \times C$$
$$= 2 \times 10^3 \times 5000 \times 10^{-6}$$
$$= 2 \times 5000 \times 10^{-3}$$
$$= 10\ 000 \times 10^{-3}$$
$$= 10\ S$$

where: R = resistance (Ω)
C = capacitance (F)

Time to reach steady state conditions = 5T
$$5T = 5 \times 10\ s$$
$$= 50\ s$$

Example 2
An R.C. timing circuit, in an electronic timer, has a capacitance of 100 μF. If the timing circuit is to have a time constant of 10 s, what value of resistor is required in the circuit?

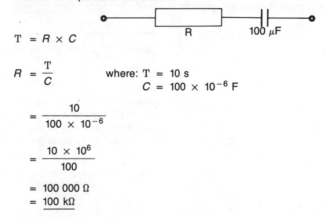

$$T = R \times C$$

$$R = \frac{T}{C}$$

where: T = 10 s
C = 100×10^{-6} F

$$= \frac{10}{100 \times 10^{-6}}$$

$$= \frac{10 \times 10^6}{100}$$

$$= 100\ 000\ \Omega$$
$$= 100\ k\Omega$$

Fig. 7.11 Time constant

100 times per second. The brightness with which the lamp glows will depend upon the charging current and this, in turn will depend upon the capacitance of the capacitor until it reaches the normal maximum brightness for the applied potential. The bigger the capacitance of the capacitor, the brighter the lamp glows. The ability of a capacitor to hold back the flow of current in an a.c. circuit is referred to as its *reactance*. Unlike the resistance of a resistor, which is independent of frequency, the reactance (X_c) of a capacitor is dependent upon frequency as the example in Fig. 7.12 shows.

7.7 Inductors and transformers

When a current flows through a coil it produces a magnetic field. At the moment of switch-on or switch-off of a direct current this magnetic field is changing very rapidly. When any magnetic field linked with a conductor changes, an electro-motive force (e.m.f.) is induced in that conductor, and if the conductor forms part of a closed circuit an induced current will flow. This induced current always flows in the

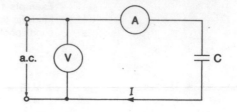

(a) An a.c. capacitive circuit

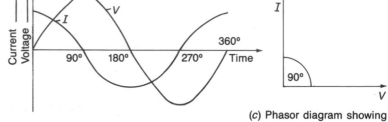

(b) Voltage and current in an a.c. capacitive circuit

(c) Phasor diagram showing that capacitive current leads the voltage by 90°

$$X_C = \frac{1}{2\pi \, fC}$$

where: X_C = Capacitive reactance (Ω)
f = Frequency (Hz)
C = Capacitance (F)

Example
A 20 μF capacitor is connected across a 240 V, 50 Hz supply. Calculate:
(i) The capacitive reactance
(ii) The current flowing in the circuit

(i) $X_C = \dfrac{1}{2\pi fc}$

$= \dfrac{1}{2\pi 50 \times 20 \times 10^{-6}}$

$= \dfrac{10^6}{2\pi 50 \times 20}$

$= \dfrac{10^6}{2000\,\pi}$

$= 159\ \Omega$

where: f = 50 Hz
C = 20×10^{-6}F
X_C = Capacitive reactance

Note
If the frequency was doubled to 100Hz the reactance (X_c) would fall to 79.5Ω.
That is, the reactance varies inversely with change of frequency

(ii) $X_C = \dfrac{V}{I}$

$I = \dfrac{V}{X_C}$

$= \dfrac{240}{159}$

$= 1.5\ A$

where: V_C = 240 V
X_C = 159 Ω
I = current

Fig. 7.12 Capacitive reactance

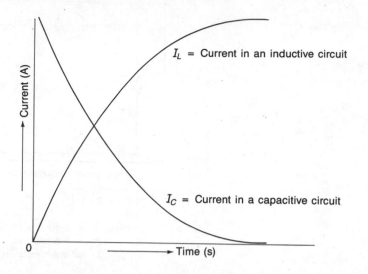

I_L = Current in an inductive circuit

I_C = Current in a capacitive circuit

Fig. 7.13 Comparison of inductor and capacitor current curves (d.c.)

opposite direction to the current creating the magnetic field in the first place, (Lenz law). Thus the applied current is held back by the induced current until the magnetic field becomes constant. This is shown in Fig. 7.13, which also compares the current flow in an inductor with that in a capacitor. It can be seen that they show opposite effects. If an alternating current is applied to the inductor, the magnetic field is constantly changing so that the applied current is constantly held back by the induced current and the current flow through the inductor is limited. Inductors are also called *chokes*. An inductor or choke is used to limit the flow of current through a fluorescent light, after it has been automatically switched to provide the initial, starting, voltage surge.

Inductors are wound from wire and even the most conductive wire has a natural resistance. Thus it is impossible to produce an inductor which has pure inductance and no resistance. There is no circuit symbol for combining the effects of inductance and resistance and the inductor and its associated resistance have to be shown separately. An inductor in series with a d.c. circuit only offers the natural resistance of the wire from which it is wound to the flow of current under steady state conditions. However, as previously stated, it offers not only its resistance but also its *reactance* to an alternating current. Unlike resistance which is independent of frequency, reactance is dependent upon the frequency of the current. Figure 7.14 shows how reactance is calculated for a pure inductor. Note that the formula used is different to that used for the calculation of capacitive reactance. Figure 7.15 shows how the *impedance* is calculated for a practical inductor, taking into account both resistance and reactance.

Inductors (chokes) vary in construction depending upon the inductance required, the frequency at which they have to operate and the application. Chokes operating at low frequencies will have many turns

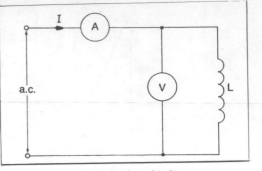

An a.c. purely inductive circuit

$$X_L = 2\pi fL$$

where: X_L = Inductive reactance
      f = Frequency (Hz)
      L = Inductance (H)

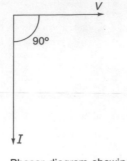

Phasor diagram showing that inductive current lags the voltage by 90°. (Conventional rotation of the phasors is anticlockwise).

Example
An inductance of 2 H is connected across a 50 Hz a.c. supply. Calculate:

(i) The reactance (X_L) of the inductor
(ii) The p.d. across the inductor if a current of 0.5 A flows through it.

(i) $X_L = 2\pi fL$ where: f = 50 Hz
       $= 2\pi \times 50 \times 2$ L = 2 H
       $= 628\ \Omega$

(ii) $V_L = I.X_L$ where: I = 0.5 A
       $= 0.5 \times 628$ X_L = 628 Ω (from (i))
       $= 314$ V

Fig. 7.14 Inductive reactance

of wire wound on a massive iron core, whilst chokes operating at very high radio frequencies consist of only a few spaced turns of silver-plated copper wire in such a manner as to be air-cored and self-supporting.

The effects described so far are referred to as *self-inductance*. When coils are allowed to interact with each other the effect is referred to as *mutual-inductance* or 'transformer effect'. A typical transformer circuit diagram is shown in Fig. 7.16. The coils are wound on a common soft-iron core to ensure maximum transfer of the magnetic flux field between the coils. The coil connected to the alternating current source is called the *primary winding*, and the coil in which the e.m.f. is induced and to which the load is connected is called the *secondary winding*. The magnetic flux in the core is created by passing an alternating current through the primary winding. This magnetic flux will constantly change in sympathy with the changing current. Since the changing flux is linked with the secondary winding, an e.m.f. will be induced in this coil. The relationship between the potential (volts)

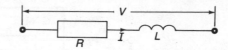

A practical inductor combines both reactance (X_L) and resistance (R)

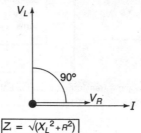

Phasor diagram for a practical inductor showing the inductive voltage (V_L) leading the current by 90°, whilst the resistive voltage (V_R) is in phase with the current.
Note (i) Voltage leading is the same as current lagging (Fig. 7.14).
(ii) Current is constant in a series circuit and is the datum phasor.

$$Z = \sqrt{(X_L^2 + R^2)}$$

Where: Z = the combined impedance
X_L = the inductive reactance
R = the resistance of the windings

The impedance triangle

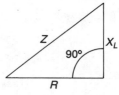

Example
Calculate the overall impedance of the 2 H inductor in the previous example if the resistance of its windings is 50 Ω.

$$Z = \sqrt{(X_L^2 + R^2)}$$
$$= \sqrt{(628^2 + 50^2)}$$
$$= \sqrt{(396\ 884)}$$
$$= \underline{630\ \Omega}$$

where: X_L = 628 Ω (Fig. 7.14)
R = 50 Ω

Note: that, unlike resistors in series, the vector sum of the reactance and resistance is less than the arithmetic addition.

Fig. 7.15 Impedance

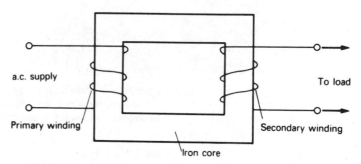

a.c. supply

To load

Primary winding

Secondary winding

Iron core

Fig. 7.16 Voltage transformer circuit diagram

across each winding and the number of 'turns' of the coils for each winding is shown in the worked example in Fig. 7.17.

Transformers are very efficient and the transfer of power is better than 90% in a well made example. The design of transformers varies depending upon their application from those with heavy iron cores for low frequency power supplies to small air-cored or ferrite dust-cored examples for radio frequency use.

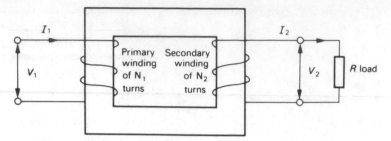

When a transformer is on load, the terminal p.d. of the
secondary winding falls to V_2

$$\frac{V_1}{V_2} = \frac{N_1}{N_2}$$ where: V_1 = p.d. across primary winding
V_2 = p.d. across secondary winding
N_1 = number of turns on primary winding
N_2 = number of turns on secondary winding

Example
A transformer takes a current of 5 A when it is connected to a
240 volt supply. The primary winding has 720 turns and the
secondary winding has 360 turns. Neglecting efficiency,
calculate:

(a) The secondary p.d.
(b) The secondary current

(a) $\dfrac{V_1}{V_2} = \dfrac{N_1}{N_2}$

$\therefore V_2 = \dfrac{V_1 \, N_2}{N_1}$

where: V_1 = 240V
N_1 = 720 turns
N_2 = 360 turns
V_2 = secondary p.d.

$= \dfrac{240 \times 360}{720}$

$= \underline{120 \text{ V}}$

Note: Had the transformer been less than 100% efficient, V_2
would have remained the same but I_2 would have been reduced
by the amount related to the drop in efficiency.

(b) $I_1 \, N_1 = I_2 \, N_2$

$\therefore I_2 = \dfrac{I_1 \, N_1}{N_2}$

where: I_1 = 5 A
N_1 = 720 turns
N_2 = 360 turns
I_2 = secondary current

$= \dfrac{5 \times 720}{360}$

Fig. 7.17 Transformer calculations

$= \underline{10 \text{ A}}$

7.8 Semiconductor (solid state) materials

As their name implies, semiconductor materials only contain a limited number of free and mobile negatively charged particles (electrons) compared with a conductor material such as copper. One of the best known semiconductor materials is *silicon* from which 'silicon chips' are made. Semiconductor materials suitable for making solid state devices are capable of having their conduction properties changed during manufacture.

Semiconductor materials contain two kinds of charge carrier. In a pure (*intrinsic*) semiconductor material such as silicon there are four outer electrons to each atom and all of them are required to form bonds with adjoining silicon atoms so that no electrons are 'spare' or free for conduction. However, as the temperature of the silicon rises the thermal energy causes the atoms to vibrate so that some of the outer electrons break free. As the temperature continues to rise, more and more electrons break free so that the conductivity of semiconductor materials improves the hotter they become. For example, the *thermistor* (Section 7.4) is made from pure silicon and its resistance to the passage of an electric current falls dramatically (conduction improves) as its temperature rises. Since the atoms of semiconductor materials are electrically neutral, every time an electron breaks free it leaves a 'hole' behind in the atomic bonding structure. This hole behaves as though it is a positive charge. Like the free negative electrons, these positive holes seem to move through the semiconductor material and form part of the electric current in it. Conduction of electricity through a semiconductor by means of the movement of thermally generated electrons to positive holes (formed in pairs) is called *intrinsic conduction* and it is controlled solely by the temperature of the pure (intrinsic) semiconductor material.

To make practical semiconductor devices the pure (intrinsic) semiconductor material is *doped* with impurities during manufacture. Semiconductor materials are said to be *tetravalent*, that is, they contain four electrons in the outer electron shells of their atoms. If atoms containing five electrons in their outer shells (such as phosphorus, antimony or arsenic) are added to the crystal, spare electrons will be available for conduction of electricity. These impurities containing five electrons are said to be *pentavalent* and conduction by their spare electrons is called *extrinsic conduction*. Since these pentavalent impurities provide negative electrons as the charge carriers, semiconductor materials containing such impurities are called *n-type* materials.

If the semiconductor material is doped with a trivalent impurity (contains 3 electrons in the outer shell of each atom), there will be an electron missing in its bonding structure for each impurity atom added. Suitable impurity materials are aluminium, indium, or gallium. With electrons missing, the charge carriers are 'holes' (positive charges) and semiconductor material with extra positive (hole) charge carriers are called *p-type* materials.

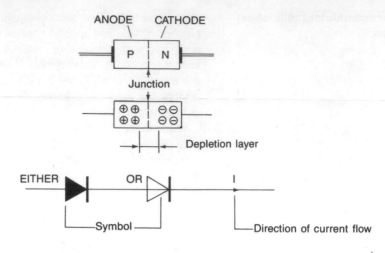

(a) **The junction diode**

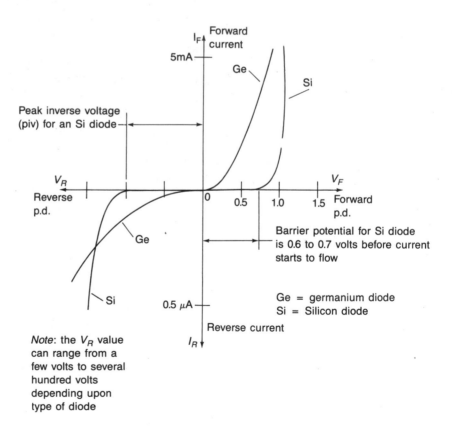

Note: the V_R value
can range from a
few volts to several
hundred volts
depending upon
type of diode

Fig. 7.18 Junction diode   (b) **Diode characteristics**

7.9   Semiconductor diodes

The most commonly used diode is the p-n junction diode. This consists of a piece of intrinsic silicon material which has been doped to give a material with n-type characteristics for half its length and to give a material with p-type characteristics for the other half of its length. Where the two regions meet is called the *junction*, hence the name 'junction diode'. This is shown in Fig. 7.18(*a*) and its characteristics are shown in Fig. 7.18(*b*). Where the holes and electrons meet at the junction they neutralise the charge on each other to form a *depletion layer* in which there are *no* free electrons and the resulting *barrier potential* must be overcome by the forward p.d. before the diode will conduct as shown in Fig. 7.18(*b*). Although Fig. 7.18(*b*) shows a higher barrier potential for the silicon (Si) diode, its steeper V_F/I_F curve and its greater peak inverse voltage (p.i.v.) makes it superior to the germanium (Ge) diode for power handling applications. On the other hand, the lower barrier potential of the germanium (Ge) diode (about 200 mV) makes it more responsive and suitable for small signal radio frequency (RF) applications.

Figure 7.19 shows two ways of connecting such a diode. In Fig. 7.19(*a*) the lamp will light. This is because the diode has not only been *biased in the forward direction*, but because this bias is sufficient to overcome the 'barrier potential' of the depletion layer. This causes the p-type charges and n-type charges to cross the junction and an electric current to flow. In Fig. 7.19(*b*) the lamp will not light because *reverse bias* has been applied and this has resulted in the depletion layer widening so that no current flows. Thus the diode can be considered as an electronic switch which will only allow current to flow in one direction.

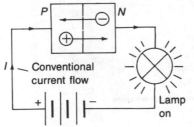

'Holes' and electrons cross the boundary layer and current flows

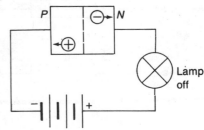

Reverse bias widens the depletion (boundary) layer and no current flows

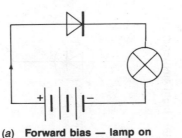

(*a*)   **Forward bias — lamp on**

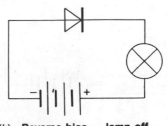

(*b*)   **Reverse bias — lamp off**

Fig. 7.19   Diode — forward and reverse bias

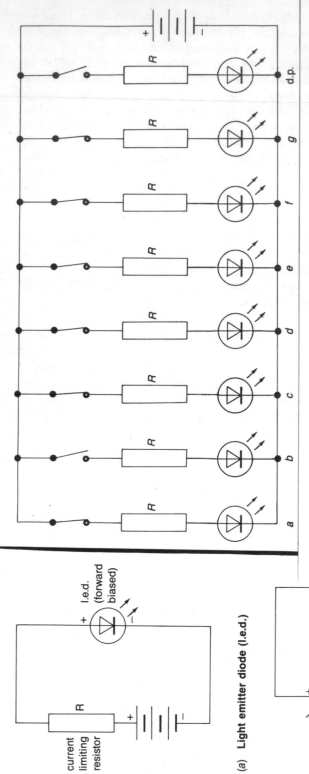

R = current limiting resistors
Segments a, c, d, e, f, g, are switched to light up. This is number 6.

(b) **7-segment l.e.d. display**

current limiting resistor

(a) **Light emitter diode (l.e.d.)**

l.e.d. (forward biased)

Photodiode reverse biased

Like all diodes, photodiodes (light sensitive junction diodes) will not normally conduct when reverse biased
However they become conductive when bright white light falls on them.

(c) **Photodiode**

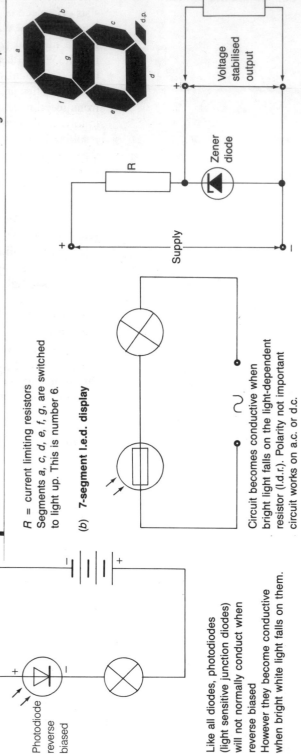

Circuit becomes conductive when bright light falls on the light-dependent resistor (l.d.r.). Polarity not important circuit works on a.c. or d.c.

(d) **Light-dependent resistor (l.d.r.)**

Voltage stabilised output

Supply

Zener diode

(e) **Zener diode**

Fig. 7.20 Miscellaneous diode types

There is one final factor to consider. Since the semiconductor material is not a perfect insulator, a very small current (a few micro-amperes) will flow in the reverse direction when reverse biased. A point is reached where, if the reverse potential is large enough, the system breaks down and a heavy reverse current called an *avalanche current* will suddenly flow and the diode may be destroyed. The highest potential which can be safely applied in the reverse direction is called the *peak inverse voltage* (p.i.v.) for the diode and this must not be exceeded.

There are a number of interesting variations on the basic junction diode and some of these will now be considered.

Light-emitting diode (l.e.d.)

These are junction diodes made from gallium phosphide and gallium arsenide and they emit light when biased in the forward direction when a current flows through them. The colour of the light depends upon the semiconductor material used. Compared with filament lamps the current to make them glow is very small and thus very little heat is emitted. To limit the current flowing through the l.e.d. to a safe value, a series resistor must always be wired into the circuit as shown in Fig. 7.20(*a*). A seven segment l.e.d. display is shown in Fig. 7.20(*b*). Each segment is a separate l.e.d. as is the decimal point (d.p.). Each segment and/or the decimal point can be made to light up in any combination to produce the number required. In the circuit shown manual switching has been used for simplicity. In practice, electronic logic switching is used.

Photodiode

The photodiode or light sensitive diode has the useful property that its reverse bias resistance falls and allows a larger reverse current to flow when light falls upon it. All semiconductor devices are sensitive to light to some extent which is why they are encased in a light-proof coating. Photodiodes are constructed to exploit this property and are encased in transparent coatings. The photodiode is shown in Fig. 7.20(*c*).

Light-dependent resistor

Although not a diode, this device is based upon semiconductor materials such as cadmium sulphide. The resistance of the device falls as the intensity of the light falling upon it increases in intensity. In fact, in some devices, the resistance can fall from as high as 10 megohms in the dark to only a few ohms in bright light. A grid of metallic electrodes is printed on the surface of a piece of cadmium sulphide. The whole device is enclosed in a case of clear resin. Not being a diode, the l.d.r. can conduct in either direction and is suitable for use in both d.c. and a.c. circuits as shown in Fig. 7.20(d).

Zener diode

This type of diode, shown in Fig. 7.20(e), is designed to be reverse biased and to run under avalanche current conditions. Providing the power rating of the zener diode is not exceeded, it will not be harmed. Reference back to Fig. 7.18(b) shows that once the 'avalanche voltage' has been reached the current flow increases rapidly, and that small changes in voltage will result in large changes in current flow through the diode. These diodes are made in a range of avalanche or zener voltages from 3 V to 150 V. Figure 7.20(e) shows a typical circuit using a zener diode as a voltage stabilising device. Any fall in current through the load R_L results in a voltage rise across the diode. This, in turn, causes an increased current to flow through the diode and this compensates for the lower load current. Thus the current through the resistor R is maintained constant and, as a result, the potential across the zener diode will remain essentially constant. In theory there will have to be small voltage changes to promote the current changes necessary to compensate for the change in load current. But as the characteristic curve in Fig. 7.18(b) shows, the current-to-voltage ratio after breakdown is so large that, for all practical purposes, the voltage across the diode can be considered constant. Conversely, any rise in current through the load results in a corresponding fall in current through the diode, and again this keeps the potential across the load constant. The same arguments can be applied to fluctuations in the supply voltage. If this falls, the voltage across the load resistor tends to fall, therefore its current tends to fall and the zener diode corrects the situation by drawing less current and allowing the controlled voltage to rise to the stabilised level. This can only happen providing the supply voltage does not fall below the avalanche voltage.

7.10 Transistors

Figure 7.21(a) shows the circuit symbol and construction of an n-p-n bipolar transistor, whilst Fig. 7.21(b) shows the construction of a p-n-p bipolar transistor. In both cases the transistor consists of layers of n-type and p-type semiconductor material. Wires are connected to each layer, and these layers are called the *emitter, base* and *collector*. The 'emitter' emits (sends) charge carriers through the thin 'base' layer to be collected by the 'collector'.

In an n-p-n type transistor the emitter sends electrons through the base to the collector, whilst in a p-n-p transistor the emitter sends positively charged 'holes' through the base to the collector. In both cases the arrowhead on the emitter shows the direction of conventional (positive) current flow.

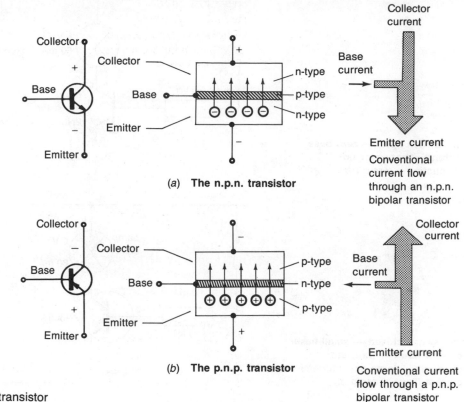

Fig. 7.21   The bipolar transistor

One of the uses of a transistor is as a switching device with a small current in the base circuit controlling a large current in the emitter-collector circuit. This is the basis of the d.c. amplifier. Figure 7.22(a) shows a simple switching circuit. When the base of the transistor is connected to the 6.0 V supply the transistor emitter-collector circuit conducts and the lamp lights. When the base is disconnected the transistor ceases to conduct and the lamp goes out. In this circuit a base current of approximately 1.0 mA is controlling an emitter-collector current of some 60 mA. This is a current gain (amplification) of 60.

$$\text{Current amplification } (\beta) = \frac{I_C}{I_B}$$

However the base current can also be made to control the magnitude of the emitter-collector current, so that a fluctuating current in the base circuit is copied and amplified in the emitter-collector circuit. This is the basis of the a.c. amplifier used for sound reproduction and a typical circuit is shown in Fig. 7.1. Further circuits will be considered later in this chapter.

Another widely used device is the Field Effect Transistor (FET). The symbol for a Junction Unigate Field Effect Transistor (JUGFET)

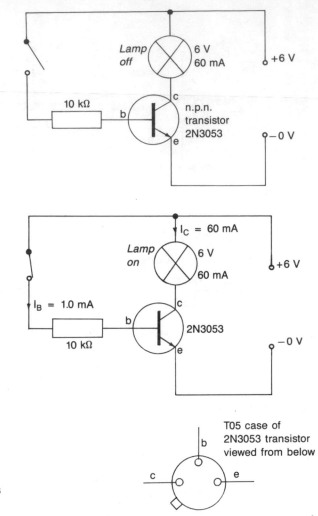

(a) **Switch open — zero base current — transistor not conducting — *lamp off***

(b) **switch closed — small base current flows — transistor conducts — large collector/emitter current flows — *lamp on***

Fig. 7.22 The bipolar transistor as an amplifier

is shown in Fig. 7.23(*a*), whilst its construction is shown in Fig. 7.23(*b*). This is only one of many different types of field effect transistors. Figure 7.24 shows a FET in a simple amplifier circuit. Unlike the bipolar transistor, negligible current flows in the control electrode (gate) of a FET. This results in the input of a FET having a very high impedance so that it does not load the input signal source. Since negligible current flows in the gate circuit and the FET only responds to gate potential, it is used as a *voltage amplifier*, unlike the bipolar transistor which behaves as a *current amplifying* device with low input and output impedances compared with the FET. Modern practice tends to combine many transistors and their associated circuits onto a single slice of silicon (a chip) to form an integrated circuit (i.c.). Circuits using integrated circuit devices can be found later in this chapter.

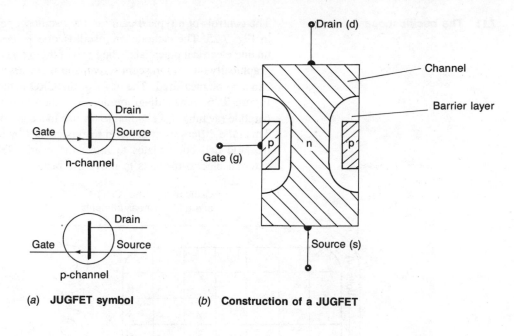

(a) **JUGFET symbol**   (b) **Construction of a JUGFET**

Fig. 7.23 Field effect transistor

Note: there are many other types of field effect transistors (FET) for special applications.

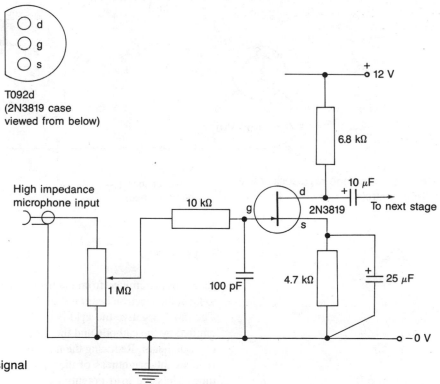

Fig. 7.24 FET as a small signal amplifier

7.11 The oscilloscope

The controls of a typical cathode ray oscilloscope (CRO) are shown in Fig. 7.25. The cathode ray oscilloscope is used to examine fluctuating electrical potentials. The form of these potentials is reproduced graphically on a fluorescent screen and their magnitude and duration can also be measured. The heart of the CRO is the cathode ray tube. Figure 7.26 shows, diagrammatically, the construction of a typical cathode ray tube (CRT) suitable for use in a cathode ray oscilloscope. This tube differs from those in television sets by using internal electrostatic deflection plates to move the beam. Television tubes use external electro-magnets to move the beam.

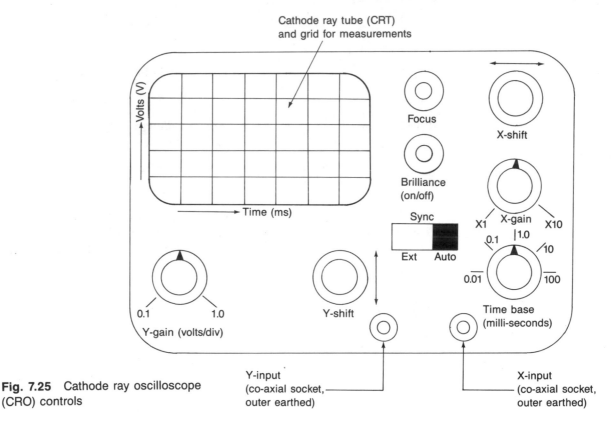

Fig. 7.25 Cathode ray oscilloscope (CRO) controls

Brightness

Electrons are emitted from the heated cathode, and the number allowed to form the electron beam is controlled by the negatively charged *grid*. The more negative the grid is made, the more it repels the electrons emitted by the cathode and the fewer electrons find their way into the electron beam. Reducing the number of electrons in the electron beam reduces the brightness of the 'trace' on the screen at the end of the tube. Thus the grid potential is used to control the brightness.

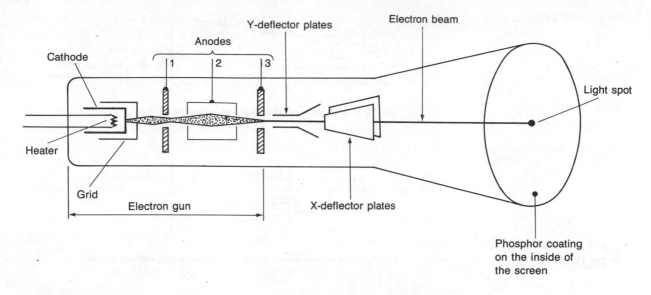

Fig. 7.26 Cathode ray rube (CRT)

Focus

The anodes used to accelerate the electrons and focus them into a narrow beam are labelled 1, 2 and 3 in the diagram. Anodes 1 and 3 are discs with a central hole in them (rather like washers) for the electron beam to pass through. Anode 2 is cylindrical and is used to focus the beam on the fluorescent screen. The anodes become increasingly positive compared with the cathode in order from 1 to 3 in the diagram. Focusing is achieved by varying the potential of the cylindrical anode. The cathode, grid and anodes form the *electron gun*.

Deflection plates

The action of the deflection plates is shown in Fig. 7.27. If the Y_1 plate is connected to the positive terminal of a d.c. supply and the Y_2 plate is connected to the negative terminal of that d.c. supply, the negatively charged electron beam will be deflected towards the Y_1 plate as shown in Fig. 7.27(*a*) and the light spot will move to the top of the screen. (Like charges repel: unlike charges attract.) If the polarity of the plates is reversed, the beam will be deflected towards the Y_2 plate as shown in Fig. 7.27(*b*) and the light spot will move to the bottom of the screen. If the Y plates are connected to an alternating source of e.m.f., the electron beam will swing up and down rapidly and continuously. It will appear to produce a vertical line on the screen. This line is called a *trace*.

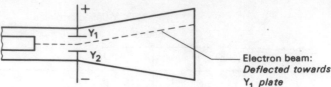

(a) **Y₁ Plate positive and Y₂ plate negative**

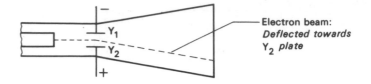

Fig. 7.27 Deflection of the electron beam (Y-plates)

(b) **Polarity of deflector plates reversed with respect to *(a)***

Time-base

In order that the trace shall represent the waveform of the applied e.m.f. it is necessary to move the beam sideways as well as up and down. To do this, a potential is applied to the X plates (see Fig. 7.28(*a*)). This potential, which must change proportionally with time, is generated by the *time-base*. The trace moves from left to right as X_2 is made increasingly positive with respect to X_1. When the trace has reached the limit of its travel the time-base output potential drops suddenly to zero and the trace returns to the left-hand end of the screen again. This is called 'fly-back' and the electron beam is usually suppressed (turned off) during fly-back. Thus the time base is a highly stable 'saw-tooth' oscillator and its output potential resembles the waveform shown in Fig. 7.28(*b*). If an a.c. supply is connected to a cathode-ray oscilloscope and the time-base is correctly adjusted, the waveform traced on the screen will resemble that shown in Fig. 7.28(*c*).

7.12 The cathode ray oscilloscope as a voltmeter

The screen of the cathode ray oscilloscope is fitted with a transparent grid so that measurements of the trace can be made. To measure a d.c. potential the following procedure is used.

(*a*) Switch off the time-base.
(*b*) Switch the input to d.c.
(*c*) Establish the zero volt datum by shorting the Y input terminals and, using the X and Y shift controls, centre the spot on the grid.
(*d*) Remove the shorting link and apply the d.c. potential to the Y plates. The spot will move up or down depending upon the polarity of the input.
(*e*) Note the number of divisions and parts of a division that the spot

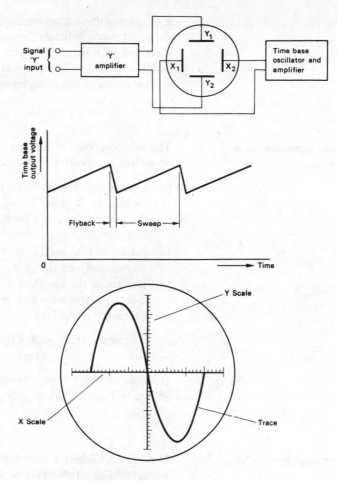

(a) **Connection of the time base**

(b) **Waveform of time base output**

(c) **Oscilloscope display of an alternating current sine waveform**

Fig. 7.28 Deflection of the electron beam (X-plates)

has moved and calculate the potential (volts) as follows:

d.c. voltage = displacement of spot (divs) × volts/division (Y-gain setting).

To measure an a.c. potential the following procedure is used.

(a) Switch off the time-base.
(b) Switch the input to a.c.
(c) Connect the a.c. potential across the Y plates.
(d) Use the X and Y shift controls to position the vertical trace conveniently on the grid.

Fig. 3.60

(e) Note the number of divisions and parts of a division that the vertical trace occupies and calculate the potential (volts) as follows:

peak voltage = 0.5 × vertical height of trace × volts/division (Y-gain setting).

peak-to-peak voltage = vertical height of trace × volts/division (Y-gain setting).

Examples of these measurements are shown in Fig. 7.29(a). Note that the oscilloscope indicates the peak a.c. voltage and not the r.m.s. value indicated by a meter. Further, because the input impedance of the oscilloscope is very high, typically several megohms, it places virtually no load on the circuit being tested, thus there is no meter insertion loss.

7.13 The oscilloscope as a frequency meter

The oscilloscope can also be used to measure the frequency of a waveform irrespective of its shape, using the following procedure.

(a) Connect the source to be measured to the Y plates.
(b) Switch the X plates to the time-base.
(c) Set the time-base to a convenient value to show rather more than one complete cycle of the waveform.
(d) Adjust the X and Y shift controls to position the waveform conveniently on the grid.
(e) Determine the length of one complete waveform (one complete cycle) in divisions and part of a division, and calculate the frequency as follows.

Frequency (Hz) = 1000/(length of one waveform (divs) × time base setting (ms/div)).

Examples for sine wave, triangular-wave and square-wave forms are shown in Figs 7.29(b) to 7.29(d).

7.14 Power supply circuits

Having established a knowledge of a range of components and an understanding of the two most important measuring instruments, namely the multi-range test meter (Chapter 6) and the oscilloscope (Section 7.13), it is now possible to consider some typical electronic circuits and their performance.

The first circuit to be considered is the mains operated power supply. This consists of four elements.

(a) A transformer to reduce the mains potential to that required by the circuit to be powered. Usually, about 12 volts.
(b) A rectifier circuit for converting the low voltage a.c. into d.c.
(c) A smoothing circuit to remove the ripple of the rectified a.c. in order to make it suitable for powering electronic devices.
(d) A voltage stabilising circuit for applications where the load current swings between wide limits, or where the application is sensitive to fluctuations in supply voltage. For example oscillator circuits (Section 7.16) are susceptible to 'frequency drift' if their supply voltage fluctuates.

Figure 7.30 shows three rectifier circuits and the corresponding output waveform as it would appear on an oscilloscope.

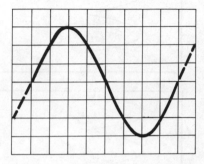

Y-gain setting	Voltage
(i) 5 V/div	No continuous trace, therefore d.c.
	Upward movement of spot therefore positive.
	$V = 3 \text{ div} \times 5 \text{ V/div}$
	$\quad = 15 \text{ volts}$
(ii) 1V/div	continuous trace therefore a.c.
	Peak voltage $= 0.5 \times 4 \text{ div} \times 1 \text{ V/div}$
	$\qquad\qquad\quad = 2 \text{ volts}$
	Peak to Peak voltage $= 4 \text{ volts}$

(a) **Oscilloscope as a voltmeter**

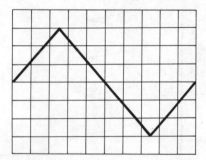

Time-base setting	Frequency
2 ms/div	$f = \dfrac{1000}{l(\text{divs}) \times \text{t.b.(ms/div)}}$
	$\quad = \dfrac{1000}{8 \times 2}$
	$\quad = 62.5 \text{ Hz}$

(b) **Frequency of a sine-wave form**

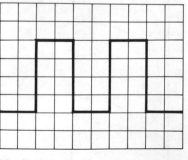

0.5 ms/div	$f = \dfrac{1000}{l(\text{divs}) \times \text{t.b.(ms/div)}}$
	$\quad = \dfrac{1000}{10 \times 0.5}$
	$\quad = 200 \text{ Hz}$

(c) **Frequency of a triangular-wave form**

5 ms/div	$f = \dfrac{1000}{l(\text{divs}) \times \text{t.b.(ms/div)}}$
	$\quad = \dfrac{1000}{4 \times 5}$
	$\quad = 50 \text{ Hz}$

(d) **Frequency of a square-wave form**

Fig. 7.29 Oscilloscope as a measuring instrument

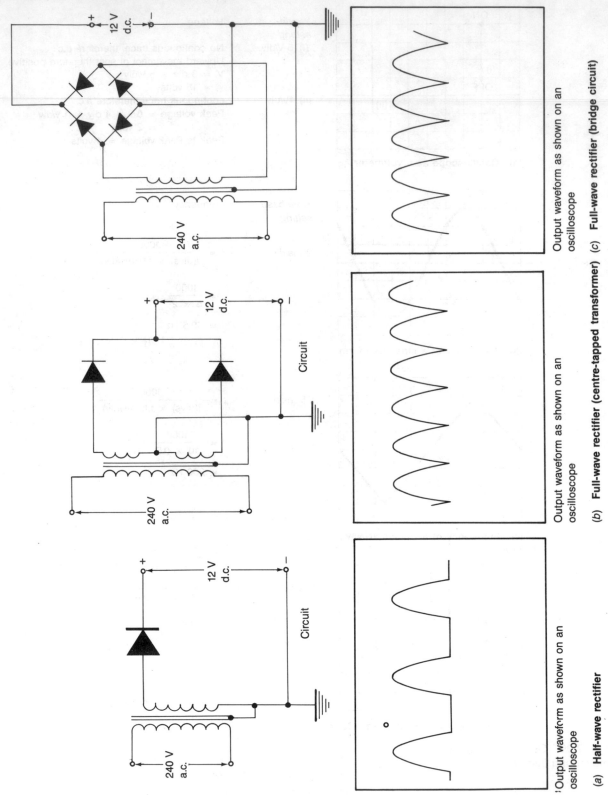

Output waveform as shown on an oscilloscope   (c) Full-wave rectifier (bridge circuit)

Output waveform as shown on an oscilloscope   (b) Full-wave rectifier (centre-tapped transformer)

Output waveform as shown on an oscilloscope   (a) Half-wave rectifier

Fig. 7.30 Rectification

Half-wave rectifier

This uses a single diode to provide the rectification after transformation of the supply potential down to the required value. The diode only conducts in the 'forward' direction, so only the positive pulses of the a.c. waveform appear in the output. The diode suppresses the negative half cycle. This would appear on an oscilloscope as indicated below the circuit. Since only half the available waveform is used this is not only inefficient, but it is very difficult to smooth the output and remove the ripple. This type of rectifier circuit would only be used where simplicity and low cost is important and where the loading on the circuit is low.

Full-wave rectifier (centre-tapped transformer)

This circuit uses a special transformer with a centre-tapped secondary winding. It also uses two diodes arranged as shown. It conducts in a forward direction twice each cycle as the alternating current swings about the centre tap of the transformer which provided the zero reference potential. Thus there are no 'gaps' in the output waveform which is shown below the circuit. This type of rectifier is much more efficient and is more suitable where substantial output currents are required. It is also much easier to smooth the output. This type of circuit was widely used in the days when only thermionic valve diodes were available. These were expensive compared with solid state diodes and difficult to connect into a bridge circuit, thus it was more economical to use a transformer with a centre-tapped secondary winding and only two diodes in a more simple circuit.

Full-wave rectifier (bridge circuit)

This circuit is the most widely used today. It avoids the need for a special and expensive transformer. Further, although four diodes are used, this is still economical as solid state diodes are relatively inexpensive. Again there are two forward pulses during each cycle resulting in easy smoothing, good power handling capacity, and high efficiency.

Figure 7.31 shows three ways of smoothing the output from the rectifier. A single high capacity, electrolytic capacitor can provide a reasonable degree of smoothing as shown. The amount of residual ripple will depend upon:

(a) whether the rectification is half-wave or full-wave;
(b) the capacitance of the capacitor;
(c) the load applied to the circuit.

Furthermore, if the capacitor is large enough to provide adequate smoothing under load conditions, the current surge each time the capacitor charges up is only limited by the resistance of the circuit

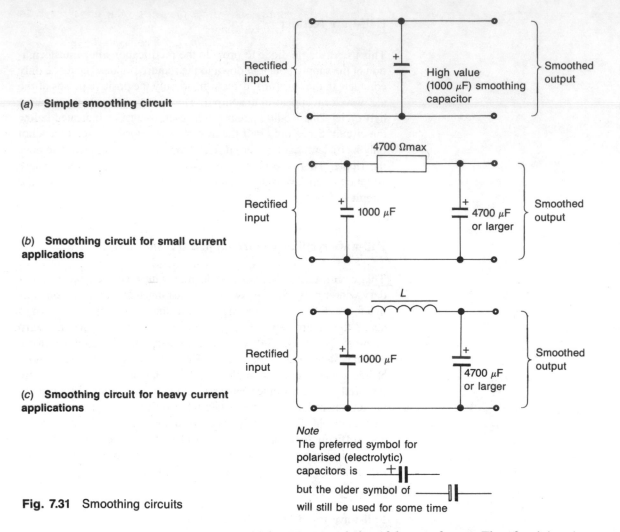

(a) **Simple smoothing circuit**

(b) **Smoothing circuit for small current applications**

(c) **Smoothing circuit for heavy current applications**

Note
The preferred symbol for polarised (electrolytic) capacitors is

but the older symbol of

will still be used for some time

Fig. 7.31 Smoothing circuits

and the output regulation of the transformer. Therefore it is quite possible to overload the diode(s) and cause early failure.

The use of two capacitors and a resistor or two capacitors and a choke is much more satisfactory. Not only is the smoothing more effective, but the resistor or choke limits the charging surge of the second capacitor which can, therefore, be made very large. The use of a resistor is satisfactory only when the load current is small, as in a radio receiver. High power amplifiers and radio transmitters demanding heavy currents from the power supply normally use the choke type filter. The choke has a low d.c. resistance and is wound on an iron core. The d.c. volt drop across it is small and its high inductance not only improves the smoothing but also improves the regulation of the power supply.

The potential at the output of the smoothing filter is greater than that measured at the output of the transformer. This is due to the fact that the smoothing capacitors charge up to the *peak* voltage of the rectified waveform and this is 1.4 times the r.m.s. value measured at the transformer secondary winding.

Many applications require *regulated* (stabilised) voltage outputs, that is, the output voltage remains constant irrespective of fluctuations in supply voltage and load current. A simple circuit using a zener diode has already been considered in Section 7.9. A more sophisticated circuit is shown in Fig. 7.32(*a*). Since the transistor TR1 is in series with the supply to the load it must be capable of carrying the full load current. In high power circuits several transistors may have to be connected in parallel to carry the load current. The transistor TR2 acts as a d.c. amplifier and controls TR1. The zener diode D_z provides a constant voltage reference. Should the output potential at *A* rise due to a reduction in load current or a rise in supply voltage, the potential at *B* will also rise. Since the potential across the zener diode will remain constant, any increase in potential at *B* will result in an increase in the base-emitter potential of TR2. This will result in an increase in collector current. Since this is being drawn through the resistor R_1, there will be a corresponding drop in collector potential for TR2. This, in turn, will increase the base-emitter potential of TR1, thus increasing its effective resistance and reducing the output potential at *A*. The small-value capacitors C_2 and C_4 prevent instability due to 'hunting' if the load is constantly varying.

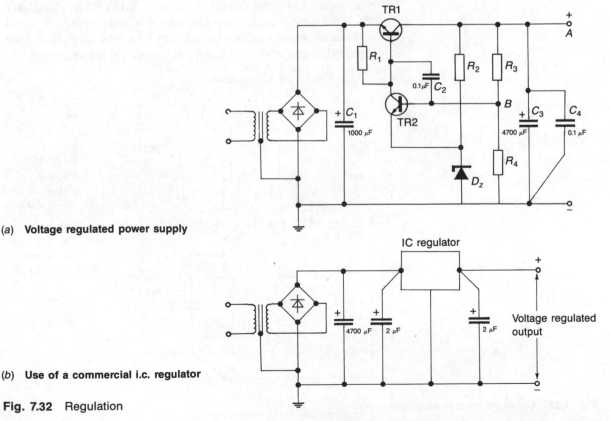

(*a*) **Voltage regulated power supply**

(*b*) **Use of a commercial i.c. regulator**

Fig. 7.32 Regulation

Should the output potential at *A* fall then the effect will be the reverse of that described above and the effective resistance of TR1 will fall and the output potential will rise. For most applications it is now normal practice to use one of the commercially available integrated circuit regulators as shown in Fig. 7.32(*b*). These are not only more sophisticated than the simple circuit just described, but they also have a degree of circuit protection built in. They are relatively cheap and can handle currents up to 10 amperes. For higher currents several such devices may be connected in parallel. The simplicity of the external circuit is apparent when compared with that in Fig. 7.32(*a*). The 2 μF capacitors must be connected close to the pins of the regulator to prevent instability. The use of a regulator also simplifies the smoothing filter when only small or medium currents are being drawn.

Note: Power supply units operate at mains voltage and great care must be employed in their construction and testing to avoid fire and shocks which could be lethal.

7.15 Amplifiers

Figure 7.33 shows a typical common emitter transistor amplifier. This is a linear or 'class-A' amplifier. The term 'linear amplifier' is derived from the characteristic curve for a transistor wired in this configuration. Figure 7.34 shows that the characteristic curve is like a long letter 'S'. If the output signal of the amplifier is to reflect the input signal without any distortion, then the amplification must take place at the straight line part of the curve (*AB*). Hence the name *linear amplifier*.

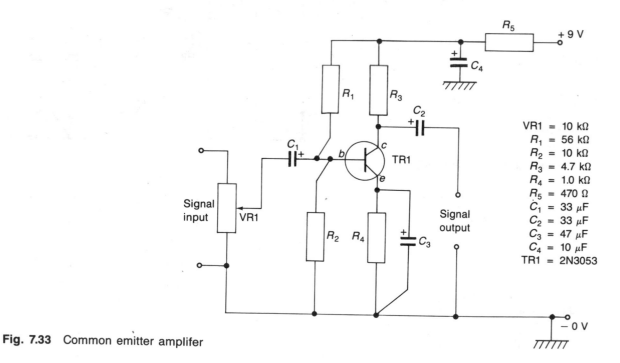

VR1	= 10 kΩ
R_1	= 56 kΩ
R_2	= 10 kΩ
R_3	= 4.7 kΩ
R_4	= 1.0 kΩ
R_5	= 470 Ω
C_1	= 33 μF
C_2	= 33 μF
C_3	= 47 μF
C_4	= 10 μF
TR1	= 2N3053

Fig. 7.33 Common emitter amplifer

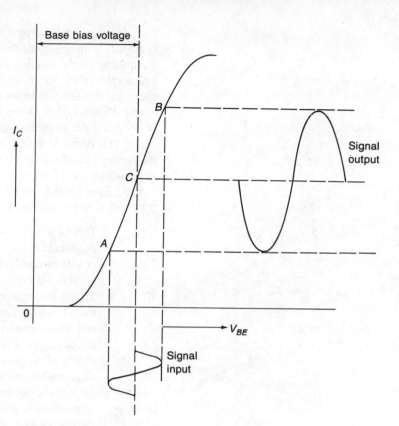

Base bias voltage

B

Signal output

C

A

I_C

0

V_{BE}

Signal input

(a) **Class A (linear) operation**

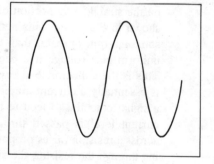

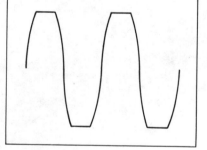

Signal waveform (output) under true class A conditions

Output 'clipped' due to overloading the amplifier resulting from a reduction in collector voltage or too large a signal input of both. This results in considerable distortion and the generation of unwanted harmonics.

(b) **Distortion in a class A (linear) amplifier**

Input signal is a pure sine-wave in both cases.

Fig. 7.34 Characteristics of a linear amplifier

The advantages of this type of amplifier are its simplicity and its ability to amplify with negligible distortion of the signal waveform. Its main disadvantage is the fact that a constant collector-emitter current flows irrespective of the signal level. Thus the class A amplifier has a low operating efficiency in terms of the power drawn for a given level of output. Figure 7.34(b) shows the effect on the output signal waveform of operating the amplifier beyond the limits of the linear part of the curve. This occurs if the d.c. supply voltage to the transistor falls (i.e. the battery runs down) or if it is driven too hard (the volume control is turned up too high), or a combination of both conditions.

The purpose of the various components in the circuit shown in Fig. 7.33 will now be described.

VR1	This is a potentiometer arranged so that it controls the magnitude of the incoming signal. It is referred to as a gain or volume control because any variation in input signal level is reflected correspondingly in the output signal level.
C_1	This is the input coupling or blocking capacitor. It allows the a.c. signal current to reach the base of the transistor and, thus, couples the amplifier to the previous stage. At the same time it blocks any d.c. component from the power supply of the previous stage reaching the base of the transistor and upsetting its bias. Hence the alternative names for this component. It makes no difference which is used.
R_1 & R_2	These form a potential divider to apply *bias* potential to the base of the transistor and ensure that the amplifier works on the straight-line portion of the characteristic curve as shown in Fig. 7.34. This network should pass a current some ten times greater than the base current to ensure a uniform bias voltage irrespective of the signal level.
R_3	This is the collector load resistor. The bipolar transistor is essentially a current amplifier and in order to achieve a signal potential to feed to the next stage, the changing current level is passed through R_3. Since the potential across a resistor varies proportionally to the current passing through the resistor, the signal potential across R_3 will be proportional to the amplified current.
C_2	This is the output blocking or coupling capacitor and is similar in function to C_1. It allows the a.c. signal current to be passed forward to the next stage of the amplifier, whilst blocking any d.c. component from the power supply to this amplifier stage.
R_4	This resistor provides protection to the transistor against excess current failure due to overheating (thermal 'runaway'). When any solid state device starts to heat up, its resistance rapidly falls. Transistors are no exception and, since they are not perfect conductors, some heat will be generated internally due to the passage of collector-emitter current. This rise in temperature results in a decrease in

internal resistance, a corresponding increase in collector-emitter current and a further rise in temperature. This process continues until the heat generated by the increased current destroys the transistor. The resistor R_4 prevents this happening. Any increase in current through the transistor, and thus through R_4, results in a corresponding increase in potential across R_4. This results in the emitter becoming more positive with respect to the base, which is the same as the base becoming more negative with respect to the emitter and the collector emitter current is reduced. This fall in the current passed by the transistor cooling down until the system becomes stabilised.

C_3 If this capacitor is omitted, an out-of-phase signal appears across R_4 and this is the equivalent of applying 'negative feedback' to the amplifier, resulting in a loss of amplification, C_3 prevents this happening by allowing the d.c. control voltage to develop across R_4 whilst, at the same time, bypassing any a.c. signal component to earth. C_3 is called a *decoupling* capacitor and, in an audio frequency amplifier, it has a value of about 50 microfarads.

R_5 & C_4 These components form a decoupling network. Without them, fluctuations of collector current in sympathy with the amplified signal would be fed back into the 9 volt d.c. supply rail. The amplified signal would then be fed back to the input via the previous stages and the amplifier would burst into spontaneous oscillation. Every stage of an amplifier other than the output stage needs to be decoupled from the d.c. supply rail.

This amplifier circuit can be tested by means of a signal generator and two oscilloscopes as shown in Fig. 7.35. This signal generator, as its name implies, is a variable frequency a.c. source. It should be switched to give a suitable level of signal of sine-wave form. In place of the two oscilloscopes shown, a single dual-beam instrument can be used so that both the input and output signals appear one above the other on the same screen. The input and output voltage levels can be measured to determine the gain of the amplifier. At the same time the output waveform can be checked for clipping and distortion. It should be an enlarged replica of the input waveform.

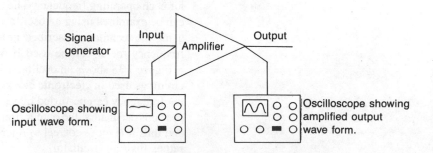

Fig. 7.35 Testing an amplifier

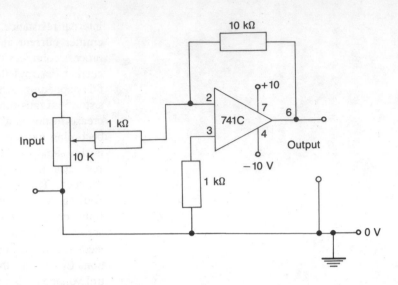

Fig. 7.36 An integrated circuit amplifier

Figure 7.36 shows an amplifier using a 742C integrated circuit. This not only gives more gain than the single transistor amplifier shown in Fig. 7.33, but the external circuitry is much simpler, having fewer components. The performance of this amplifier can be checked using a signal generator and oscilloscopes as previously described. Note that a dual power supply is required giving d.c. potentials which are positive and negative about the zero point.

7.16 Oscillators

Oscillators are amplifier circuits in which positive feedback is deliberately introduced at a level which will cause instability and spontaneous oscillation. Figure 7.37 shows two such oscillator circuits. That shown in Fig. 7.37(*a*) is a Colpitts-type electron coupled oscillator (ECO), whilst that shown in Fig. 7.37(*b*) is a Hartley-type inductively coupled oscillator. Because they oscillate spontaneously, they are called 'free-running' oscillators. The frequency of oscillation will depend upon the circuit constants and, particularly, the values of *L* and *C* which form a parallel tuned circuit. Note the use of a variable capacitor to vary the frequency of oscillation. Quartz crystals also act as very efficient and stable tuned circuits and Fig. 7.37(*c*) shows a crystal controlled version of the ECO. The frequency stability of a crystal tuned oscillator is of a very high order, but a different crystal is required for each operating frequency. The output waveform of these oscillators can be examined using an oscilloscope and their frequency measured using the technique described in Section 7.13. Alternatively a digital frequency meter can be used if one is available.

Figure 7.38 shows an oscillator based upon the type 555 integrated circuit as used in electronic clocks. This is a highly stable device and the frequency can be varied by varying the values of VR1. Unlike the previous oscillators, the output from this circuit has a 'square' waveform and thus it is more correct to refer to this circuit as a *pulse generator* rather than an oscillator.

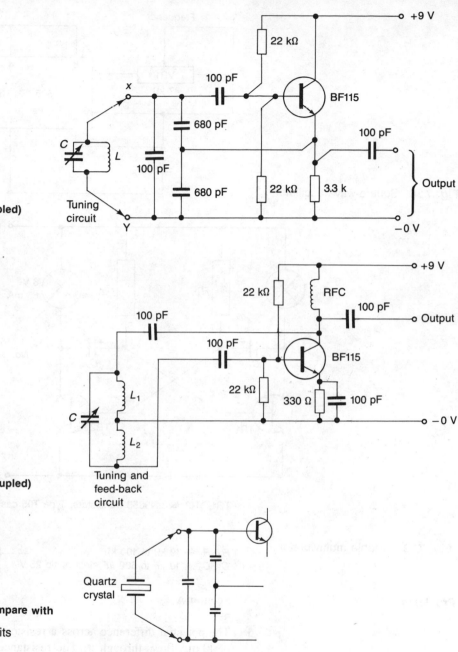

(a) **Colpitts (electron-coupled) oscillator (ECO)**

(b) **Hartley (inductively-coupled) oscillator**

(c) **Crystal tuned ECO (compare with**

Fig. 7.37 Oscillator circuits

The last circuit to be considered in this section is the *astable multivibrator* or 'flip-flop' oscillator (see Fig. 7.39). Increasing the values of C_1, R_1 and C_2, R_2 slows down the rate at which the lamps flash on and off. (Remember: time constant = CR.) The lamps should flash on and off alternately and for equal periods. It is interesting to observe what happens if C_1, R_1 and C_2, R_2 are not symmetrical in value.

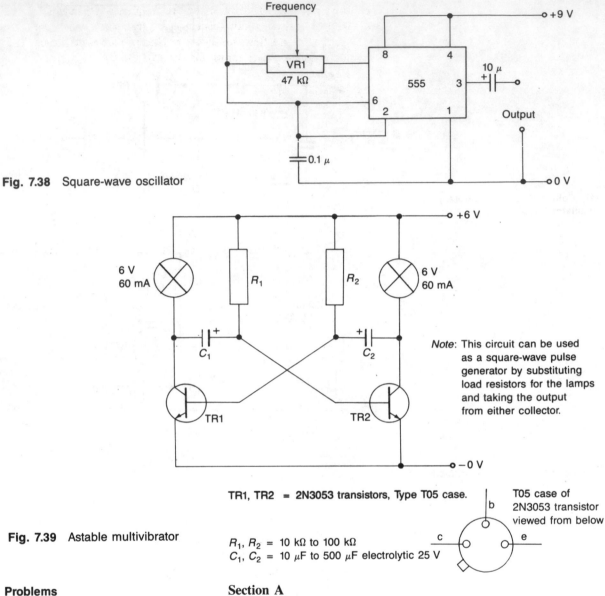

Fig. 7.38 Square-wave oscillator

Note: This circuit can be used as a square-wave pulse generator by substituting load resistors for the lamps and taking the output from either collector.

TR1, TR2 = 2N3053 transistors, Type T05 case.

T05 case of 2N3053 transistor viewed from below

Fig. 7.39 Astable multivibrator

R_1, R_2 = 10 kΩ to 100 kΩ
C_1, C_2 = 10 μF to 500 μF electrolytic 25 V

Problems

Section A

1 The potential difference across a resistor is 6 volts when a current of 10 mA flows through it. The resistance of this resistor is:
 (*a*) 0.6 Ω
 (*b*) 60 Ω
 (*c*) 600 Ω
 (*d*) 6000 Ω

2 Which of the following statements about resistance and reactance is true?
 (*a*) Resistance is frequency dependent.
 (*b*) Reactance is frequency dependent.

(c) Both resistance and reactance are frequency dependent.

(d) Neither resistance nor reactance are frequency dependent.

3   A current of 15 mA flows through a resistor of 1000 Ω resistance. The smallest suitable power rating for the resistor is:

(a) 1/8 watt;

(b) 1/4 watt;

(c) 1/2 watt;

(d) 1 watt.

4   Alternating current flowing through an inductor (choke) is opposed by its:

(a) dielectric;

(b) resistance alone;

(c) reactance alone;

(d) impedance.

5   A potential difference of 2000 volts is applied across the terminals of a capacitor of 10 μF capacitance. The charge on the capacitor in coulombs will be:

(a) 0.02 coulombs;

(b) 0.2 coulombs;

(c) 200 coulombs;

(d) 20 000 coulombs.

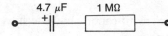

4.7 μF   1 MΩ

Fig. 7.40

6   The time constant for the circuit shown in Fig. 7.40 is:

(a) 0.47 seconds;

(b) 4.7 seconds;

(c) 470 seconds;

(d) 4700 seconds.

7   A voltage transformer has an input of 240 volts and an output of 16 volts. Given that the number of turns for the primary winding is 1680, the number of turns for the secondary winding will be:

(a) 112;

(b) 105;

(c) 85;

(d) 15.

8   The current gain (β) for a common emitter amplifier is given by the expression:

(a) $\beta = I_C \times I_B$

(b) $\beta = \dfrac{I_C}{I_B}$

(c) $\beta = I_E \times I_B$

(d) $\beta = \dfrac{I_E}{I_B}$

9   The voltage reference source in a voltage stabilised power supply is provided by a:

(a) thermistor;

(b) transistor;

(c) zener diode;

(d) quartz crystal.

10 A quartz crystal may be used in an oscillator circuit to:
    (a) control the frequency of oscillation;
    (b) control the power output;
    (c) prevent the transistor from overheating;
    (d) control the collector voltage of the transistor.

Section B

11 State one essential similarity and one essential difference between an amplifier and an oscillator circuit and sketch an example of a typical circuit for each.

12 Describe, with the aid of sketches, how an oscilloscope can be used to examine the waveform of a Colpitts-type electron coupled oscillator tuned by an LC circuit. Sketch a typical output waveform for this type of oscillator as it would appear on the screen of the oscilloscope.

13 (a) Describe with the aid of sketches how an oscilloscope can be used to measure the gain of an amplifier.
    (b) The input to the amplifier is of sine-waveform. Sketch the appearance of the output waveform if the amplifier is overloaded.

14 The power supply circuit shown in Fig. 7.41 provides full-wave rectification.
    (a) Explain what is meant by 'full-wave' rectification and state its advantages over 'half-wave' rectification.
    (b) Describe the function of the components C_1, C_2 and L, and state the main problem which would arise if their values were too low.

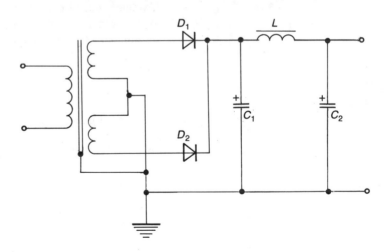

Fig. 7.41

15 Describe what is meant by the following terms as applied to amplifier circuits:
    (a) linear amplification;
    (b) positive feedback;
    (c) negative feedback;
    (d) decoupling.

16   Describe *three* advantages and *one* diadvantage of using an integrated circuit in place of discrete components when designing an electronic device.

17   Describe typical applications for the following electronic devices:
(*a*) thermistor;
(*b*) light emitter diode;
(*c*) photodiode;
(*d*) zener diode.

18   A choke has an inductance of 0.5 H and the resistance of its windings is 100 Ω. Calculate:
(*a*) its reactance for a frequency of 250 Hz;
(*b*) its impedance for a frequency of 250 Hz.

19   Figure 7.42 shows a resistance network. Calculate:
(*a*) the resistance of the circuit between the points *A* and *B*;
(*b*) the power dissipated by the circuit;
(*c*) the potential difference across the 100 Ω resistor;
(*d*) the current flowing through the 60 Ω resistor.

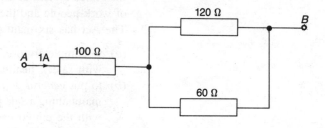

Fig. 7.42

20   (*a*) State the essential differences between *bipolar* and *field effect* transistors, and describe a typical application for both devices.
(*b*) State the essential difference between *n.p.n.* and *p.n.p.* bipolar transistors and give an example of the use of each device.

8 Safety hazards

8.1 Health and Safety at Work Act

It is essential to observe safe working practices not only at work but also during your practical assignments in the college workshops and laboratories.

The Health and Safety at Work Act provides a comprehensive, integrated system of law dealing with the health, safety and welfare of work-people and the general public as affected by work activity. The Act has six main provisions:

(a) to completely overhaul and modernise the existing law dealing with safety, health and welfare at work;
(b) to put *general duties* on employers ranging from providing and maintaining a safe place to work, to consulting on safety matters with the employees;
(c) to create a Health and Safety Commission;
(d) to reorganize and unify the various government inspectorates;
(e) to provide powers and penalties for the enforcement of safety laws;
(f) to establish new methods of accident prevention, and new ways of operating future safety regulations.

The Act now places the responsibility for safe working equally upon:

(a) the employer;
(b) the employee;
(c) the manufacturers and suppliers of goods and equipment.

8.2 Health and Safety Commission

The Act provides for a full-time, independent chairman and between six and nine part-time commissioners. The commissioners are made up of three trade union members appointed by the TUC, three management members appointed by the CBI, two Local Authority members, and one independent member.

The commission has taken over the responsibility previously held by various government departments for control of most occupational safety and health matters. The commission is also responsible for the organization and functioning of the *Health and Safety Executive*.

8.3 Health and Safety Executive

This unified inspectorate combines together the formerly independent government inspectorates such as the Factory Inspectorate, the Mines and Quarries Inspectorate, and similar bodies. Since 1975 they have been merged into one body known as the Health and Safety Executive Inspectorate. The inspectors of the HSE have wider powers under the Health and Safety at Work Act than under previous legislation and their duty is to implement the policies of the Commission.

8.4 Enforcement

Should an inspector find a contravention of one of the provisions of the existing Acts or Regulations, or a contravention of a provision of the new Act, the inspector has three possible lines of action available.

Prohibition Notice

If there is a risk of serious personal injury, the inspector can issue a Prohibition Notice. This immediately stops the activity giving rise to the risk until the remedial action specified in the notice has been taken to the inspector's satisfaction. The Prohibition Notice can be served on the persons undertaking the dangerous activity, or it can be served on the person in control of the activity at the time the notice is served.

Improvement Notice

If there is a legal contravention of any of the relevant statutory provisions, the inspector can issue an Improvement Notice. This notice requires the fault to be remedied within a specified time. It can be served on the person deemed to be contravening the legal provision, or it can be served on any person on whom responsibilities are placed. This latter person can be an employer, employee, or a supplier of equipment or materials.

Prosecution

In addition to serving a Prohibition Notice or an Improvement Notice, the inspector can prosecute any person (including an employee — *you*) contravening a relevant statutory provision. Finally, the inspector can *seize, render harmless, or destroy* any substance or article which he considers to be the cause of imminent danger or personal injury.

Thus every employee, trainee or experienced worker must ensure that he or she is a fit and trained person to carry out his or her assigned tasks. By law every employee must:

(*a*) obey all safety rules of his or her place of employment;

(*b*) understand and use, as instructed, the safety practices incorporated in particular activities or tasks;

(c) not proceed with his or her task if any safety requirement is not thoroughly understood. Guidance must be sought;

(d) maintain his or her working area tidy and maintain his or her tools in good condition;

(e) draw the attention of the safety officer or his or her immediate superior to any potential hazard;

(f) report all accidents or incidents (even if injury does not result from that incident) to the responsible person;

(g) understand emergency procedures in the event of an accident or an alarm;

(h) understand how to give the alarm in the event of an accident or an incident such as fire;

(i) cooperate promptly with the senior person in charge in the event of an accident or incident such as fire.

Therefore, safety, health and welfare are very personal matters for the young worker just entering industry. This chapter sets out to identify the main hazards and discuss how they can be avoided. Factory life, and particularly engineering, is potentially dangerous and a positive approach must be taken towards safety, health and welfare.

8.5 Accidents

Accidents do not happen, they are caused. There is not a single accident which could not have been prevented by care and forethought on somebody's part. Accidents can and must be prevented. They cost millions of pounds every year in damage and loss of premises, plant and lost business. They cost millions of lost man-hours of production each year, but even this is of little importance compared with the immeasurable cost in human suffering.

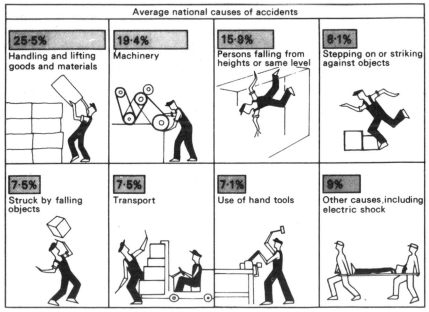

Fig. 8.1 Causes of industrial accidents

In every eight-hour shift nearly 100 workers are the victims of industrial accidents. Many of these will be blinded, maimed for life, or confined to a hospital bed for months. At least two of them will die. Figure 8.1 shows the main causes of accidents.

8.6 Protective clothing

For general workshop purposes the *boiler suit* is the most practical and the safest form of body protection. However, to be completely effective certain precautions must be taken as shown in Fig. 8.2.

Long hair

(a) Long hair is liable to be caught in moving machinery, particularly drilling machines and lathes. The resulting wound (the hair and scalp being torn away) is both extremely painful and dangerous. Brain damage may also occur.

(b) Long hair is also a health hazard, as it is almost impossible to keep it clean and free from infection in the workshop environment (see Section 8.7).

Sharp tools

Sharp tools protruding from the breast pocket can cause severe wounds to the wrist. Since the motor nerves of the fingers are near the surface in the wrist, these wounds can result in paralysis of the hand and fingers.

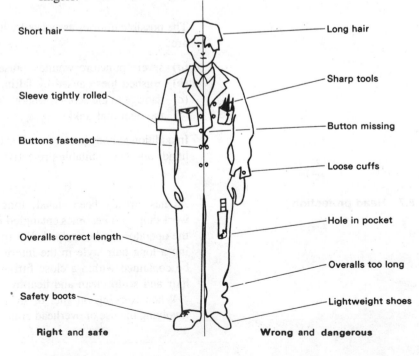

Short hair — — Long hair

Sleeve tightly rolled — — Sharp tools

Buttons fastened — — Button missing

— Loose cuffs

— Hole in pocket

Overalls correct length — — Overalls too long

Safety boots — — Lightweight shoes

Right and safe — **Wrong and dangerous**

Fig. 8.2 Correct dress (courtesy EITB)

Button missing

Since the overall cannot be fastened properly, it becomes as dangerous as any other loose clothing and liable to be caught in moving machinery.

Loose cuffs

Not only are loose cuffs liable to be caught up like any other loose clothing, they may also prevent the wearer from snatching his or her hand away from a dangerous situation.

Hole in pocket

Tools placed in the pocket can fall through onto the feet of the wearer. Although this may not seem potentially dangerous as the feet should be protected by stout shoes, nevertheless it could cause an accident by distracting attention at a crucial moment.

Overalls too long

These can cause falls, particularly when negotiating stairways.

Lightweight shoes

The possible injuries associated with lightweight and unsuitable shoes are:

(*a*) severe puncture wounds caused by treading on sharp objects,
(*b*) crushed toes caused by falling objects,
(*c*) damage to the achilles tendon due to insufficient protection around the heel and ankle.

In addition to body protection, it is necessary to protect the head, eyes, hands and feet. Suitable protective clothing will now be considered.

8.7 Head protection

As has already been stated, long hair is a serious hazard in the workshop. If it becomes entangled in a machine, as shown in Fig. 8.3, the operator can be scalped. If a fitter or machinist persists in retaining a long hair style in the interests of fashion, then the hair must be contained within a close fitting cap. This also helps to keep the hair and scalp clean and healthy.

When working on site, or in a heavy engineering erection shop involving the use of overhead cranes, all persons should wear a safety

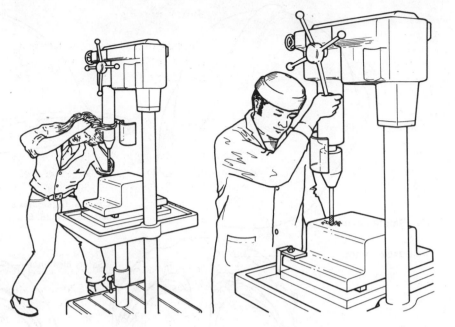

Fig. 8.3 The hazard of long hair (courtesy EITB)

helmet complying with BS 2826, as even small objects such as nuts and bolts can cause very serious head injuries when dropped from heights. Figure 8.4(*a*) shows such a helmet. Safety helmets are made from moulded plastic or from fibre-glass reinforced polyester. They are colour coded for personnel identification, and are light and comfortable to wear, yet despite their lightness they have a high resistance to impact and penetration. To eliminate the possibility of electric shock, safety helmets have no metal parts. The materials used to manufacture the outer shell have to be non-flammable and their electrical insulation must withstand 35 000 volts. Figure 8.4(*b*) shows the harness inside a safety helmet. This provides ventilation and a *fixed safety clearance* between the outer shell of the helmet and the wearer's skull. This clearance must always be maintained at 32 millimetres. The entire harness is removable for cleaning and sterilizing. It can be adjusted for size, fit and angle to suit the individual wearer's head.

8.8 Eye protection

Whilst it is possible to walk on a wooden leg, nobody has ever seen out of a glass eye. Therefore eye protection is possibly the most important safety precaution to take in the workshop. Eye protection is provided by wearing suitable goggles or visors as shown in Fig. 8.5. When welding, special goggles have to be worn with coloured lenses to filter out harmful rays. These will be referred to in detail in Section 8.21 — Oxy-acetylene (gas) welding hazards, and Section 8.22 — Arc-welding hazards. Gas welding goggles are not suitable when arc welding. Eye injuries fall into three main categories:

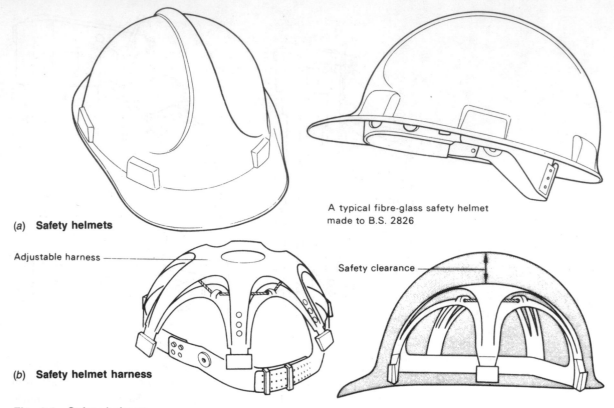

(a) **Safety helmets**

A typical fibre-glass safety helmet made to B.S. 2826

Adjustable harness

Safety clearance

(b) **Safety helmet harness**

Fig. 8.4 Safety helmets

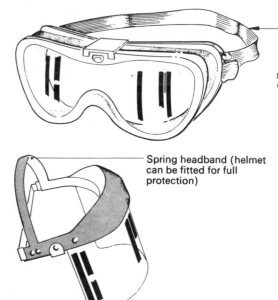

Elastic headband

Transparent plastic goggles suitable for machining operations

Spring headband (helmet can be fitted for full protection)

Complete plastic face visor for protection against chemical and salt-bath splashes

Fig. 8.5 Safety goggles and visors

(*a*) pain and inflammation due to abrasive grit and dust getting between the lid and the eye;

(*b*) damage due to exposure to ultra-violet radiation (arc-welding) and to high intensity visible radiation. Particular care is required when using laser equipment;

(*c*) loss of sight due to the eyeball being punctured or the optic nerve being severed by flying splinters of metal (swarf), or by the blast from a compressed air jet.

8.9 Hand protection

The engineer's hands are in constant use and because of this they are at risk handling dirty, oily, greasy, rough, sharp, brittle, hot and maybe toxic and corrosive materials. Gloves and 'palms' of a variety of styles and types of materials are available to protect the hands whatever the nature of the work.

Where gloves are inappropriate, as when working precision machines, and the hands need to be protected from oil and grime rather than from cuts and abrasions, then a 'barrier cream' should be rubbed in to the hands. This is a mildly antiseptic, water soluble cream which fills the pores of the skin and prevents the ingress of dirt and subsequent infection. The cream is easily removed by washing and carries away the dirt and removes sources of infection.

8.10 Foot protection

The practice of wearing unsuitable footwear should always be discouraged. It is not only false economy, but extremely dangerous to wear light-weight casual or sports shoes in the workplace. They offer no protection from 'crushing' or penetration. In safety footwear, protection is provided by a steel toe-cap (inside the boot or shoe) which conforms to a strength specification in accordance with BS 1870. Safety footwear is available in a wide range of styles and prices. It can be attractive in appearance and comfortable to wear. Figure 8.6 shows sections through safety footwear.

8.11 Health hazards

Noise

Excessive noise can be a dangerous pollutant of the working environment. The effects of noise can be:

(*a*) fatigue leading to carelessness and accidents;
(*b*) mistaken communications between workers leading to accidents;
(*c*) ear damage leading to deafness;
(*d*) permanent nervous disorders.

The level at which a noise becomes dangerous varies with its frequency band (pitch) and the length of time the worker is exposed to it. Noise is energy and it represents waste since it is useless. Ideally, it should be suppressed at source to avoid waste and to improve the working

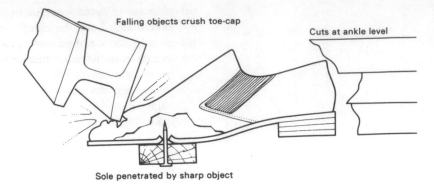

Falling objects crush toe-cap

Cuts at ankle level

Sole penetrated by sharp object

*Light-weight shoes offer **No** protection*

Steel toe-cap

Steel intersole

Non-slip oil-resistant sole

INDUSTRIAL SAFETY SHOE

Stout leather
prevents injury
to the achilles
tendon

Fig. 8.6 Safety footwear

INDUSTRIAL SAFETY BOOT

environment. If this is not possible, then the operator should be insulated from it by sound absorbant screens and/or ear-muffs.

Narcotic (anaesthetic) effects

Exposure to small concentrations of narcotic substances causes drowsiness, giddiness and headaches. Under such conditions the worker is obviously prone to accidents since his or her judgement and reactions are adversely affected. Injury can result from falls, and a worker who has become disorientated by the inhalation of narcotics is a hazard to other workers. Prolonged or frequent exposure to narcotic substances can lead to permanent damage to the brain and other organs of the body, even in relatively small concentrations. Exposure to high concentrations can result in rapid loss of consciousness and end in fatality. Examples of narcotic substances are to be found among the many solvents used in industry. Solvents are used in paint, adhesives, polishes and degreasing agents. Careful storage and use is essential and should be closely supervised. Fume extraction and adequate ventilation of the work place must be provided when working with these substances. Suitable respirators should be available for use in emergencies.

Irritant effects

Many substances cause irritation to the skin both externally and internally. They may also sensitise the skin so that it becomes irritated by substances not normally considered toxic.

External irritants can cause industrial dermatitis by coming into contact with the skin. The main irritants met with in the workshop are oils — particularly coolants — and adhesive solvents.

Internal irritants are the more dangerous as they may have deep seated effects on the major organs of the body where they may cause inflammation, ulceration, internal bleeding, poisoning and the growth of cancerous tumours. Internal irritants are usually air pollutants in the form of dusts (asbestos fibres), fumes and vapours. They may also be carried into the body on food handled without washing (see Section 8.12), or from storing noxious substances in soft-drinks bottles without proper labelling. Many domestic tragedies happen this way.

Systemic effects

Substances known as systemics affect the fundamental organs and bodily functions. They affect the heart, the brain, the liver, the kidneys, the lungs, the central nervous system and the bone marrow. Their effects cannot be reversed and thus lead to chronic ill-health and, ultimately, early death. These toxic substances may enter the body in various ways.

(a) Dust, vapour and gases can be breathed in through the nose. Observe the safety codes when working with such substances and wear the respirators provided.
(b) Liquids and powders which contaminate the hands can be transferred to the digestive system by handling food or cigarettes with dirty hands. Wash before eating.
(c) Liquids, powders, dusts, and vapours may all enter the body through the skin:
  (i) directly through the pores;
  (ii) by destroying the outer horny layers of the skin and attacking the sensitive layers underneath;
  (iii) by entering through undressed puncture wounds.

Regular washing, use of a barrier cream, use of suitable protective (rubber or plastic) gloves, immediate dressing of cuts — no matter how small — is essential.

8.12 Personal hygiene

As already stated above, personal hygiene is most important. There is nothing to be embarrassed about in rubbing a barrier cream into your hands before work, about washing thoroughly with soap and hot water after work, about changing your overalls regularly so that they can be cleaned. Personal hygiene can go a long way towards preventing skin diseases, both irritant and infectious. In some processes where gloves would hinder manual dexterity a barrier cream is the only protection available.

Skin disease due to continual contact with mineral oil forms the main health hazard in the engineering industry. The effects range from skin irritations and dermatitis to the formation of skin cancers, and will depend upon the type of oil, its temperature, and the degree and length of time of exposure. They will also depend upon the condition of the skin, cuts and abrasions from handling rough or sharp components and swarf, irritation from additives, and infection.

The first effect is usually simple irritation accompanied by redness and pimples. If treatment is not sought the condition deteriorates until cracking, scaling and skin growths appear. Even in mild cases sensitisation of the skin may occur in which case the operator may need to change to a job where oil and other irritants are not present.

Soluble oils (suds) are particularly difficult as the water content causes the skin to macerate (become soggy); this reduces its natural resistance. If excessive contact with oil cannot be reduced by modification to the plant or process, then additional water- and oil-proof protective clothing must be made available and worn. Dirty and oil impregnated overalls are also a major source of skin infection. Overalls should be changed and cleaned regularly.

8.13 Behaviour in workshops

In an industrial environment horseplay infers reckless, foolish and boisterous behaviour such as pushing, shouting, throwing things and practical joking by a person or a group of persons. Such actions can distract a worker's attention and break his or her concentration and lead to serious — even fatal — accidents. There is no place for such foolish activity in industry; yet it occurs in some way in every type of firm every day.

Horseplay observes no safety rules. It has no regard for safety equipment. It can defeat safe working procedures and undo the painstaking work of the safety officer by the sheer foolishness and thoughtlessness of the participants.

Types of accidents due to horseplay depend largely on the work of the factory concerned and the circumstances leading to the accident. Generally they are caused when a person's concentration is disturbed so that they incorrectly operate a machine or come into contact with moving machinery; when someone is pushed or knocked against moving machinery or factory transport; when they are pushed against ladders and trestles upon which people are working at heights; when they fall against and dislodge heavy stacked components, and when electricity, compressed air and dangerous chemicals are involved.

8.14 Lifting and carrying

As was shown in Fig. 8.1, the movement of materials is the biggest single cause of factory accidents. Manual handling accidents can be traced to one or more of the following:

(a) incorrect lifting technique;
(b) carrying too heavy a load;
(c) incorrect gripping;
(d) failure to wear protective clothing.

Figure 8.7(a) shows the wrong technique for lifting which can lead to ruptures, strained backs, sprains, slipped discs and other painful and permanent injuries. The correct technique is shown in Fig. 8.7(b). The back is kept straight and the lift comes from the powerful leg and thigh muscles. Figure 8.7(c) is a reminder that the load being carried must not obstruct forward vision, to avoid falls and injury.

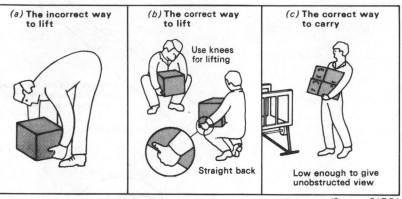

Fig. 8.7 Lifting and carrying

(Courtesy E.I.T.B.)

Protective clothing should be worn when lifting and carrying. Crushed toes caused by dropped loads can be avoided by wearing safety shoes. Cuts and splinters can be avoided by wearing suitable gloves when handling rough and sharp materials. Burns can be prevented when handling caustic and corrosive fluids by wearing face shields and rubber or plastic suits as protection against spillage and splashes. Team lifting should be employed when handling heavier loads, but remember that there can be only *one captain* to the team and only he or she gives the orders. All members of the team should lift together so as to spread the load evenly. All members of the team should be of similar physique and build.

Heavy loads should be lifted using mechanical lifting gear such as hoists, cranes, fork-lift trucks, etc. Such appliances must only be used by persons trained in their use or under the direct supervision of a person trained in their use.

8.15 Hazards associated with hand tools

The newcomer to industry often does not realise the potential danger existing in badly maintained and incorrectly used hand tools. Unfortunately the newcomer is often influenced by older men and women — who should know better — misusing hand tools, often through

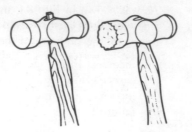

Loose hammer head and split shaft.   *Chipped, cracked and mushroomed hammer head.*

Blunt cutting edge, chipped and mushroomed head.

(a) Hammer faults

(b) Chisel faults

Strained and cracked jaws caused by extending the handle with a tube.

Tang not protected by a suitable handle.

(c) Spanner faults

(d) File faults

Hand tools in a dangerous condition

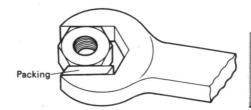

Packing

(e) Do not use an oversize spanner and packing. Use the correct size of spanner for the nut or bolt head.

(f) Do not use a file as a lever.

Fig. 8.8 Dangers in use of hand tools

Misuse of hand tools

laziness. The time and effort taken in fetching the correct tool from the stores or in servicing a worn tool is considerably less than the time taken in convalescing from an injury. Figure 8.8 shows some of the hazards arising from the incorrect use of hand tools.

8.16 Hazards associated with portable power tools

In addition to maintaining portable power operated tools in good order, it is imperative that the equipment and particularly the power lead is electrically sound or a fatal electric shock can occur to the user.

(*a*) Be sure that the tool is properly earthed or that a 'double-insulated' tool is used.

(*b*) Report the following unsafe conditions and do not use the tool until they have been put right by a qualified electrician:

(i) defective or broken insulation both to the tool and its flex;

(ii) improperly or badly made connections to terminals;

(iii) broken or defective plug;

(iv) loose or broken switch;

(v) sparking brushes.

(*c*) Do not overload a motor. The heat generated will damage the insulation.

(*d*) Do not use a portable electrical tool in the presence of flammable vapours and gases unless it is designed for such use. Far better to use a tool powered by compressed air. A spark from the switch or motor of an electrically powered tool could cause a fire or an explosion.

(*e*) Portable power tools should be operated from low-voltage, isolating transformers at all times. Normally these have an output potential of 110 volts, but on wet sites, out of doors, may be as low as 50 volts. The transformer should be fed from an ELCB (see Chapter 6).

8.17 Hazards associated with machine tools

Metal cutting machines are potentially dangerous. Tools designed to cut through solid metal will not be stopped by fragile flesh and bone.

(*a*) Before operating machinery be sure that you have been properly taught how to control it and the dangers associated with it.

(*b*) Do not operate a machine unless all guards and safety devices are in position and working correctly.

(*c*) Make sure you understand any special rules applicable to the *particular machine* you are about to use, even if you have been trained on machines in general.

(*d*) Never clean or adjust a machine whilst it is still in motion.

(*e*) Report any dangerous aspect of the machine immediately and stop working it until it has been made safe again by a qualified person.

(*f*) A machine may have to be stopped in an emergency. Learn how to make an emergency stop without having to pause and think about it.

Transmission guards

By law, no machine can be sold or hired out unless all gears, belts, shafts and couplings making up the transmission equipment are guarded so that they cannot be touched whilst in motion. Sometimes guards have to be removed to replace, adjust, or service the components they are covering. Before removing guards or covers:

(*a*) stop the machine;

(*b*) isolate the machine from its energy supply;

(c) lock the isolating switch so that it cannot be turned on again whilst you are working on the exposed equipment. Keep the key in your pocket;

(d) if it is not possible to lock the isolating switch, remove the fuses and keep them in your pocket.

If an 'interlocked' guard is removed, an electrical or a mechanical trip will stop the machine from operating. This trip is only provided in case you forget to isolate the machine and is no substitute for full isolation.

Cutter guards

The machine manufacturer does not normally provide cutter guards, because of the wide range of work a machine may have to do.

(a) It is the responsibility of the owner or the hirer of the machine to supply their own cutter guards.

(b) It is the responsibility of the operator to make sure that the guards are fitted and working correctly before operating a machine, and to use the guards as instructed. It is an offence in law for an operator to remove or tamper with the guards provided.

(c) A technician may have to fit and adjust a guard, or supervise a craftsperson fitting or adjusting a guard or other safety device for an unskilled operator. This is a great responsibility and the technician must thoroughly understand the function and correct fitting of the guard before performing such a duty.

(d) If ever you are doubtful about the adequacy of a guard, or the safety of a process, consult your safety officer immediately.

(e) Figure 8.9 shows a typical telescopic guard for a drilling machine.

Use of grinding wheels

Because of its apparent simplicity, the double-ended off-hand grinding machine comes in for more than its fair share of abuse. In fact a grinding wheel does not just 'rub off' the metal; it is a precision multi-tooth cutting tool in which each grain has a definite cutting geometry. Therefore the grinding wheel must be mounted, dressed and used correctly if it is to cut efficiently. Further, a grinding wheel which is damaged or incorrectly mounted or is unsuitable for the machine on which it is mounted can burst at speed and cause serious damage and injury. Thus the guard for a grinding wheel not only stops the operator from coming into accidental contact with the wheel, but also provides *burst containment* in case the wheel shatters in use, that is, the broken pieces of wheel are contained within the guard and are not thrown out from the machine at high speed. The tool rest must also be properly adjusted so that there is no chance of the work being

Metal band to clamp round spindle sleeve

Clamp bolt

Perspex guard to prevent operator coming into contact with drill and spindle

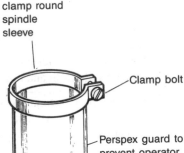

Simple drill guard

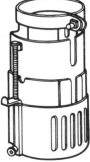

Telescopic drill guard

Fig. 8.9 Drill guards

dragged down between the tool rest and the wheel. The mounting of grinding wheels, except by a trained and registered person, is prohibited under the Abrasive Wheel Regulations, 1970.

8.18 Electrical hazards

The principles and practice of electrical installations have been introduced in Chapter 6 including overcurrent protection to prevent fire and earthing to prevent shock. The installation and maintenance of electrical equipment must be left to the specialist. However all engineers should have an understanding of the principles and practices concerned with the installation and maintenance of electrical equipment and protective devices so as to avoid misuse of the equipment and to recognize potential dangers. An electrical shock from a factory power supply can easily kill a man or woman. (Note that most machines operate at 415 volts r.m.s.) Even if the shock is not severe, the convulsion it causes can throw the victim from a ladder or against a machine and this can result in serious injury or even death. When pulling a victim clear of the fault that caused the shock be very careful, as it is possible to receive an equally severe shock from the victim. Always pull the victim by his or her clothing which, if dry, will act as an insulator. *Never touch the flesh* of the victim until he or she is clear of the fault, as flesh acts as a conductor. Artificial respiration must be commenced immediately the victim is pulled clear of the fault or the live conductor. Figure 8.10 — reproduced by courtesy of *Electrical Times* gives details of how to treat the victim of severe shock.

8.19 Fire fighting

Fire fighting is a highly skilled operation and most medium and large firms have properly trained teams who can contain the fire locally until the professional service arrives.

The best way you can help is to learn the correct fire drill; both how to give the alarm and how to leave the building. It only requires one person to panic and run in the wrong direction to cause a disaster.

In an emergency never lose your head and panic.

Smoke is the main cause of panic. It spreads quickly through a building, reducing visibility and increasing the risk of falls down stairways. It causes choking and even death by asphyxiation. Smoke is less dense near the floor: as a last resort crawl. To reduce the spread of smoke and fire, keep fire doors closed at all times but *never locked*. The plastic materials used in the finishes and furnishings of modern buildings give off highly toxic fumes. Therefore it is best to vacate the building and leave the fire-fighting to the professionals who have breathing apparatus. Saving human life is more important than saving property.

If you do have to help to fight a fire there are some basic rules to remember. A fire is the rapid oxidation (burning) of combustible (burnable) materials at relatively high temperatures. Remove the air or the fuel or lower the temperature as shown in Fig. 8.11 and the fire goes

Order of action

1 Switch off current

Do this immediately. If not possible do not waste time searching for the switch.

2 Secure release from contact

Safeguard yourself when removing casualty from contact. Stand on non-conducting material (rubber mat, DRY wood, DRY linoleum). Use rubber gloves, DRY clothing, a length of DRY rope or a length of DRY wood to pull or push the casualty away from the contact.

3 Start artificial respiration

If the casualty is not breathing artificial respiration is of extreme urgency. A few seconds delay can mean the difference between success or failure. Continue until the casualty is breathing satisfactorily or until a doctor tells you to stop.

4 Send for doctor and ambulance

Tell someone to send for a doctor and ambulance immediately and say what has happened. Do not allow the casualty to exert himself by walking until he has been seen by a doctor. If burns are present, ask someone to cover them with a dry sterile dressing.

If you have difficulty in blowing your breath into the casualty's lungs, press his head further back and pull chin further up. If you still have difficulty, check that his lips are slightly open and that the mouth is not blocked, for example, by dentures. If you still have difficulty, try the alternative method, mouth-to-mouth, or mouth-to-nose, as the case may be.

Fig. 8.10 Treatment for electric shock

Method - mouth-to-mouth

1. Lay casualty on back; if immediately possible, on a bench or table with a folded coat under shoulders to let head fall back. Kneel or stand by casualty's head. Press his head fully back with one hand and pull chin up with the other.

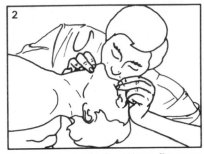

2. Breathe in deeply. Bend down, lips apart and cover casualty's mouth with your well open mouth. Pinch his nostrils with one hand. Breathe out steadily into casualty's lungs. Watch his chest rise.

3. Turn your own head away. Breathe in again.

Repeat 10 to 12 times per minute.

If the patient does not respond proceed as follows:

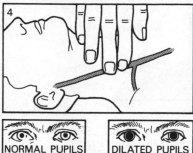

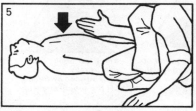

NORMAL PUPILS   DILATED PUPILS

4. Check carotid pulse, pupils of eyes and colour of skin (see Fig.4)

5. Pulse present, pupils normal - continue inflations until recovery of normal breathing (Figs 1, 2 and 3)

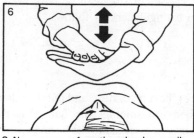

6. Pulse absent, pupils dilated, skin grey - strike smartly to the left part of breast bone with edge of hand (see Fig.5)

7. Response of continued pulse, pupils contract - continue inflations until recovery of normal breathing

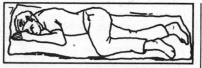

8. No response of continued pulse, pupils unaltered, skin grey - commence external heart compression (see Fig.6) *

9 When normal breathing commences, keep warm, place casualty in the recovery position (see Fig. 7)

* Method - external heart compression (Fig.6)

1. Place yourself at the side of the casualty

2. Feel for the lower half of the breastbone

3. Place the heel of your hand on this part of the bone, keeping the palm and fingers off the chest

4. Cover this hand with the heel of the other hand

5. With arms straight, rock forwards pressing down on the lower half of the breastbone (in an unconscious adult it can be pressed towards the spine for about one and a half inches (4 cm))

6. The action should be repeated about once a second

Continue as above until a continued pulse is felt and pupils contract

Continue inflations until recovery of normal breathing

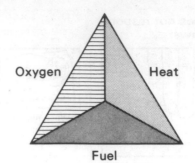

Oxygen   Heat

Fuel

These are the three 3 essentials to start a fire

Note : Once the fire has started it produces sufficient heat to maintain its own combustion reactions and sufficient surplus heat to spread the fire.

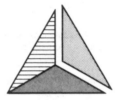  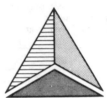

When solids are on fire remove heat by applying water.

Liquids, such as petrol etc. on fire can be extinguished by removing oxygen with a foam or dry powder extinguisher.

Electrical or gas fires can usually be extinguished by turning off the supply of energy.

(a) **Remove heat**

(b) **Remove oxygen (air)**

(c) **Remove energy source (fuel, gas, electricity etc.)**

Fig. 8.11 Fire fighting

out. It can be seen from Fig. 8.11 that different fires require to be dealt with in different ways. The main classes of fires can be correlated to the normally available portable extinguishers as follows.

Water

Used in large quantities this reduces the temperature and puts out the fire. The steam generated also helps to smother the flames as it displaces the air and therefore the oxygen essential to the burning process. However, for various technical reasons it should only be used on burning solids such as wood, paper, some plastics, etc.

Foam extinguishers

These are used for fighting oil and chemical fires and act by smothering the flames and preventing the oxygen of the air from feeding the fire. Water alone cannot be used as the burning oil floats on the surface of the water and the fire spreads.

Note: Since both water and foam are electrically conductive, do not use them on fires associated with electrical equipment or the person wielding the hose or the extinguisher could be electrocuted.

Carbon dioxide (CO_2) extinguishers

These are used on burning gases and vapours. They can also be used for oil and chemicals in confined places. The carbon dioxide gas replaces the air and smothers the fire. It can only be used in confined places where it is not dispersed by drafts.
Note: If the fire cannot breathe neither can you, so care must be taken to evacuate living creatures from the vicinity before operating the extinguisher. Back away from the bubble of CO_2 gas as you operate the extinguisher, do not advance into it.

Vaporising liquid extinguishers

These cover CTC, CBM, and BCF extinguishers.The heat from the fire causes rapid vaporisation of the liquid spray from the extinguisher and this vapour smothers the fire. It will also smother living creatures unless precautions are taken. This type of extinguisher is suitable for oil, gas, vapour and some chemical fires. Like carbon dioxide extinguishers, vaporising liquid extinguishers are safe to use on fires associated with electrical equipment.

Dry powder extinguishers

These are suitable for small fires involving flammable liquids and small quantities of solids such as paper. They are also suitable for fires in electrical equipment. The main active ingredient is powdered sodium bicarbonate (baking powder) which gives off carbon dioxide when heated. They are suitable for offices and canteens as the power leaves little mess and does not contaminate any foodstuffs with which they come into contact. The residual powder can be removed subsequently with a vacuum cleaner.

Since a fire spreads quickly, a speedy attack is essential if it is to be contained. Sound the alarm and send for assistance before attempting to fight a fire. Remember:

(*a*) extinguishers are only provided to fight small fires;
(*b*) take up a position between the fire and the exit, so that your escape cannot be cut off;
(*c*) DO NOT continue to fight the fire if:
　　(i) it is dangerous to do so;
　　(ii) there is a possibility that the escape route may be cut off by fire, smoke or collapse of the building;

(iii) the fire spreads despite your efforts;

(iv) toxic fumes are being generated by the burning of plastic finishes or furnishings;

(v) there are gas cylinders or explosive substances in the vicinity of the fire.

If you have to withdraw, close windows and doors behind you wherever possible.

Finally, ensure that extinguishers are recharged immediately after use.

8.20 Fire prevention

Prevention is always better than cure, and fire prevention is better than fire fighting. Tidiness is of paramount importance in reducing outbreaks of fire. Fires have small beginnings, and it is usually amongst rubbish left lying around that many fires originate. So make a practice of constantly removing rubbish, shavings, offcuts, cans and bottles, waste paper, oily rags, and other unwanted materials to a safe place at regular intervals. Discarded foam plastic packing, now widely used, is not only flammable, but gives off highly dangerous toxic fumes when burnt.

Highly flammable materials should be stored in specially designed and equipped compounds away from the main working areas. Only minimum quantities of such materials should be allowed in the workshop at a time, and then only into *non-smoking* zones. The advice of the Local Authority Fire Prevention Officer should be sought.

It is good practice to provide metal containers for rubbish, preferably with air-tight hinged lids, and with proper markings as to the type of rubbish they should contain since some types of rubbish *will ignite spontaneously* when mixed. The lids of the bins should be kept shut so that if a fire starts it will soon use up the limited air available and go out of its own accord without doing any damage.

Liquid petroleum gases (LPG) such as propane and butane are being used increasingly for process heating and for space heating in workshops and on site. Full and empty cylinders should be stored separately in isolated positions away from the working areas and shielded from the sun's rays. There should be plenty of air around them with free circulation and ventilation. They should be protected from frost. Where large cylinders or spheres are installed in fixed positions for the bulk storage of such gases, not only must the above precautions be observed, but they must be securely fenced and defended against damage from passing vehicles. Pipe runs, joints and fittings associated with an LPG must be regularly inspected by a qualified person, and any flexible tubing used in connection with gas cylinders should be regularly inspected for cuts and abrasions and replaced as necessary. The storage and use of oxygen and acetylene cylinders as used for welding require special treatment and are dealt with in Section 8.21.

8.21 Gas-welding hazards

The hazards associated with *oxy-acetylene welding* can be summarized as follows.

(*a*) *Eye injuries* resulting from the glare of the incandescent (white-hot) weld pool and splatter from molten droplets of metal. Eye injuries can be prevented by the use of proper welding goggles as shown in Fig. 8.12. It is essential to use goggles with the correct filter glasses depending upon the materials being joined and any flux being used, as these affect the radiation. Note that gas welding goggles offer no protection from the radiations associated with arc welding.

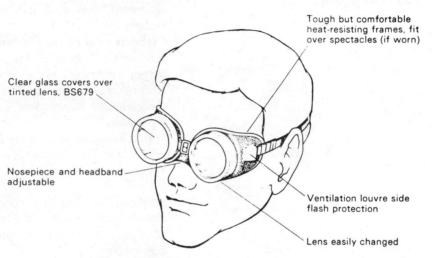

Clear glass covers over tinted lens, BS679

Tough but comfortable heat-resisting frames, fit over spectacles (if worn)

Nosepiece and headband adjustable

Ventilation louvre side flash protection

Lens easily changed

Fig. 8.12 Oxy-acetylene welding goggles

Note: GOGGLES WITH LENSES SPECIFIED FOR USE WHEN GAS WELDING OR CUTTING MUST NOT BE USED FOR ARC WELDING OPERATIONS

(*b*) *Burns* and fire hazards resulting from careless use of the torch and the careless handling of hot metal. Sparks are also a hazard. Ensure all rubbish and flammable materials are removed from the working area. Fire-resistant overalls should be worn and cuffs avoided as these present traps for sparks and globules of hot metal. For the same reason the overalls should be fastened at the neck. A leather apron should be worn together with leather spats and, in very hot conditions, gloves.

(*c*) *Explosions* resulting from the improper storage and use of compressed gases and gas mixtures. Explosions can occur when acetylene gas is present in the air in any proportion between 2% and 82%. It will also explode spontaneously at high pressure even without air or oxygen being present. The working pressure for acetylene should not exceed 620 millibars. Explosions in the equipment itself may result from *flashbacks* due to improper use or lighting up procedures, incorrect setting of the equipment, faulty equipment. For this reason flashback arrestors must be fitted and regularly inspected. An exploding acetylene cylinder is equivalent to a large bomb. It can demolish a building and kill the occupants. Great care must be taken in their use.

Gas cylinders are themselves not dangerous as they are regularly inspected to government standards. However, the following safety precautions should be observed.

(a) Cylinders must be protected from mechanical damage during storage, transportation and use. *Acetylene cylinders must always be kept upright.*

(b) Cylinders must be kept cool. On no account should the welding flame be allowed to play on the cylinder or any other part of the equipment. They must also be protected from extremes of hot sunlight and frost.

(c) Cylinders must be sited in well-ventilated surroundings to prevent the build up of explosive mixtures of gases. Even slight oxygen enrichment can cause spontaneous ignition of clothing and other flammable materials leading to fatal burns. As stated earlier concentrations of acetylene as low as 2% in air can result in explosions.

(d) Correct automatic pressure regulators must be fitted to all cylinders prior to use. The cylinder valve must always be kept closed when the equipment is not in use or whilst changing cylinders or equipment.

(e) Keep cylinders free from contamination. Oils and greases can explode spontaneously in the presence of pure oxygen. Similarly, oil or dirty cloths can ignite spontaneously in an oxygen-enriched atmosphere.

The above notes only cover the barest outlines of the safety precautions which must be observed when gas welding. Not only the safe handling of cylinders and gases must be studied and understood, but also the correct procedures for lighting up and shutting down welding equipment. The British Oxygen Company issues a number of booklets on welding safety and these should be studied and fully understood before handling gas welding equipment.

8.22 Arc-welding hazards

The hazards arising from the use of mains-operated welding equipment are set out in Table 8.1. To eliminate these hazards as far as possible, the following precautions should be taken. These are only the basic precautions and a check should be made as to whether the equipment and working conditions require special additional precautions.

(a) Make sure that the equipment is fed from the mains via an isolating switch-fuse and earth leakage circuit breaker (ELCB) (see Chapter 6). Easy access to this switch must be provided at all times.

(b) Make sure that the trailing, high-voltage, primary cable is armoured against mechanical damage as well as being heavily insulated against a supply potential of 415 volts.

Table 8.1 Arc-welding electrical hazards

CIRCUIT — HIGH VOLTAGE — Primary

Fault:	*Hazard:*
1. Damaged insulation	Fire — loss of life and damage to property Shock — severe burns and loss of life
2. Oversize fuses	Overheating — damage to equipment and fire
3. Lack of adequate earthing	Shock — if fault develops — severe burns and loss of life

---

CIRCUIT — LOW VOLTAGE — Secondary (very heavy current)

Fault:	*Hazard:*
1. Lack of welding earth	Shock — if a fault develops — severe burns and loss of life
2. Welding cable — damaged insulation	Local arcing between cable and any adjacent metalwork at earth potential causing Fire
3. Welding cable — inadequate capacity	Overheating leading to damaged insulation and Fire
4. Inadequate connections	Overheating — severe burns — Fire
5. Inadequate return path	Current leakage through surrounding metalwork — overheating — Fire

---

(c) Make sure that all cable insulation and armouring is undamaged and in good condition and that all terminations and connecting plugs and sockets are also secure and undamaged. If in doubt, do not operate the equipment until it has been checked and made safe by a skilled electrician.

(d) Make sure that all the equipment and the work is adequately earthed with conductors capable of carrying the heavy currents used in welding (see Chapter 6).

(e) Make sure that the welding current regulator has an 'OFF' position so that in the event of an accident the welding current can be stopped without having to trace the primary cable back to the isolating switch.

(f) Make sure that the 'external welding circuit' is adequate for the heavy currents it has to carry.

For all arc-welding operations it is essential to protect the welder's head, face and eyes from radiation, spatter and hot slag, and for this purpose either a helmet or hand shield is used. Examples are shown in Fig. 8.13. The injurious effect of the radiations emitted by an electric arc are similar whether a.c. or d.c. is used for welding. Exposing the head, face and eyes to infra-red (heat) rays would lead to the welder becoming uncomfortably hot and would induce serious eye troubles. If too much ultra-violet radiation is received by the welder or anyone

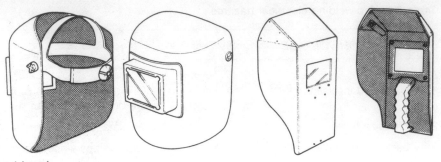

Fig. 8.13 Arc-welding eye and head protection

Arc-welder's helmet

(b) Arc-welder's hand shield

else in the vicinity, it can cause effects similar to sunburn, and a condition known as 'arc-eye'. In addition, too much visible light from the incandescent weld pool will dazzle the operator, whilst too little can cause eyestrain and headaches.

The obvious precaution is to prevent these harmful radiations from the welding-arc and from the molten weld-pool from reaching unprotected skin and from reaching the eyes. In the latter case special glass filters of suitable colour and density are used in the face mask or helmet. These not only absorb the harmful radiations but also reduce the intensity of the visible light. The expensive filter glass is protected by a cover glass on the outside. This cover glass is clear and toughened.

The slag left by the flux used when arc-welding has to be chipped away when the weld has cooled down. Clear goggles should be worn whilst chipping as the slag breaks away in glass-like splinters. Protective screens should be provided so that adjacent workers are not put at risk.

The welder's body and clothing must be protected from radiation and burns. Figure 8.14 shows a welder wearing full protective clothing whilst working overhead.

People working in the vicinity of a welding-arc, including other welders, can be exposed to stray radiations from the arc, and can be caused considerable discomfort. Wherever possible each arc should be screened in such a way as to keep stray radiations to a minimum either by the use of individual cubicles or portable screens. Looking at an unscreened welding arc, even from a distance of several metres and for only a few seconds, can cause 'arc-eye'. The painful effects of exposure will not be felt until between four and twelve hours later. The symptoms of 'arc-eye' include a feeling of 'sand in the eyes', soreness, burning and watering. If a person is exposed to a flash, the effects of 'arc-eye' can be minimized by the immediate use of a special lotion which should be available in the first-aid box.

Although arc-welding electrodes are flux coated, the welder is not likely to suffer any ill effects from welding fumes provided that reasonable ventilation is available. Localised ventilation can be provided by a suction fume extractor, which not only dilutes and removes

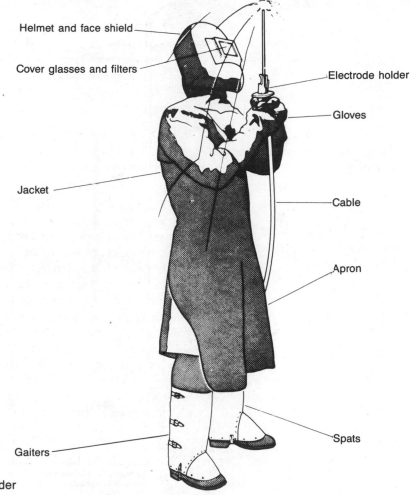

Helmet and face shield

Cover glasses and filters

Electrode holder

Gloves

Jacket

Cable

Apron

Spats

Gaiters

Fig. 8.14 Fully protected arc welder

fumes but assists in keeping down the temperature and adds to the comfort and efficiency of the welder. Extraction should be at 'low-level' so that the fumes are not drawn up past the face of the welder.

8.23 Workshop layout

Much can be done to prevent accidents by the layout of the machines and equipment in a workshop. Figure 8.15 shows a well laid-out workshop. It can be seen that there are ample gangways and that these are clearly marked and left free of obstructions. The machines are arranged so that bar stock does not protrude into the gangways. They are also arranged so that the operators' attention is not distracted by other workers constantly passing by close to them. Grinding machines are arranged so that grit is not thrown towards other machines and so that dust extraction out of the working area can easily be provided. There is easy access to emergency stop switches for the whole shop

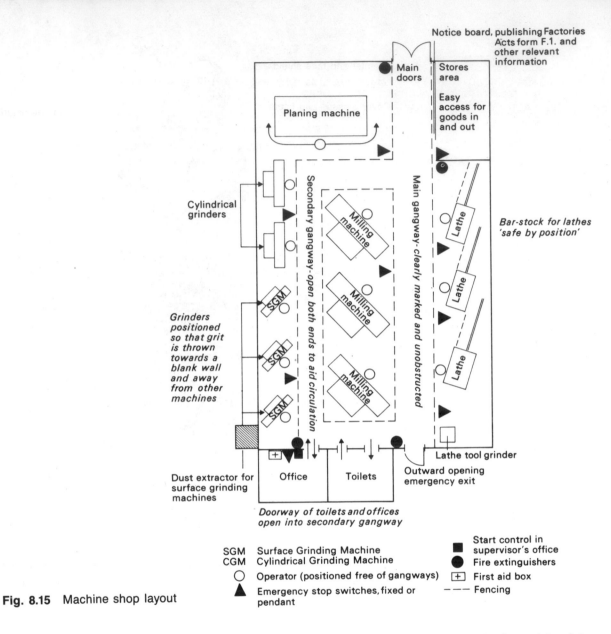

Fig. 8.15 Machine shop layout

and fire extinguishers are strategically placed. Scale models of the more common types and makes of machine tools are available, and these can be used in conjunction with scale plans of the workshop floor for experimenting with different layouts in order to achieve an efficient and safe workshop.

Problems

Section A

1 Goggles are worn when working with machine tools to:
   (*a*) improve vision;

(b) prevent glare;

(c) protect your eyes;

(d) reduce eye fatigue.

2   A guard is provided on a drilling machine to:
 (a) protect the drill from damage;
 (b) prevent coolant being thrown out by the drill;
 (c) prevent swarf being thrown out by the drill;
 (d) prevent the operator from coming into contact with the drill.

3   The most suitable type of starter switch for a machine tool is operated by a:
 (a) mushroom headed push button;
 (b) shrouded (recessed) push button;
 (c) rotary knob;
 (d) tumbler knob.

4   The manufacturer of a machine tool is responsible for providing:
 (a) transmission guards only;
 (b) cutter guards only;
 (c) both transmission and cutter guards;
 (d) neither transmission nor cutter guards.

5   When lifting heavy loads you should:
 (a) bend your back and your knees;
 (b) keep your back and knees straight;
 (c) keep your knees straight and bend your back;
 (d) keep your back straight and bend your knees.

Section B

6   Describe in detail six checks that should be made to ensure that a portable power tool is electrically safe to use.

7   Describe what precautions should be taken to ensure a reasonable level of personal hygiene when working in an engineering workshop. State *three* harmful effects that may result from not taking reasonable precautions to maintain personal cleanliness.

8   Explain why the following basic protective clothing should be worn when working in a machine shop:
 (a) boiler suit type overall;
 (b) reinforced protective shoes;
 (c) safety goggles.

9   (a) What type of fire extinguisher should be used when fighting: (i) burning oil; (ii) burning paper and rag; (iii) an electrical fire?
 (b) What action should you take if you discover a fire has started in a building where you are working?

10   A workmate has been rendered unconscious by a severe electric shock. Describe in reasonable detail how you would:
 (a) separate the victim from the source of the shock;
 (b) render artificial respiration.

Index